THE INTELLECTUAL AND SOCIAL ORGANIZATION OF THE SCIENCES

THE INTELLECTUAL AND SOCIAL ORGANIZATION OF THE SCIENCES

BY

RICHARD WHITLEY

CLARENDON PRESS - OXFORD
1984

Oxford University Press, Walton Street, Oxford OX2 6DP

London New York Toronto
Delhi Bombay Calcutta Madras Karachi
Kuala Lumpur Singapore Hong Kong Tokyo
Nairobi Dar es Salaam Cape Town
Melbourne Auckland
and associated companies in
Beirut Berlin Ibadan Mexico City Nicosia

Oxford is a trade mark of Oxford University Press

Published in the United States
by Oxford University Press, New York

British Library Cataloguing in Publication Data
Whitley, Richard
The intellectual and social organization
of the sciences.
1. Science—Social aspects
I. Title
306'.45   Q175.5
ISBN 0-19-827248-0

Set by Taj Services Ltd., New Delhi, India
and printed in Great Britain by
Billing & Sons Ltd, Worcester

For
Barbara

ACKNOWLEDGEMENTS

I have benefited greatly from the comments and suggestions of Rod Coombs, Peter Halfpenny, Jon Harwood, Chunglin Kwa, Trevor Pinch, Bob Scholte, and Terry Shinn who have read parts of this work. Many of the ideas discussed here have been presented in lectures and seminars given at the Universities of Amsterdam, Aston, Brunel, Manchester, Oxford, the Copenhagen School of Economics, the 1981 Annual Meeting of the French Sociological Society, and a 1982 British Social Science Research Council conference on Science Studies and I am grateful to the audiences on these occasions for their interest and suggestions. Much of the comparative research on the organization of scientific work in different fields, which heightened my conviction that the comparative understanding of diffeent sciences is critical to the sociology of scientific knowledge, was conducted with Allan Bitz, Penelope Frost, and Andrew McAlpine who contributed to my appreciation of organizational contingencies and differences in the production of knowledges. Finally, I am greatly indebted to Lesley Wiszniewski who has ably coped with the many revisions of this book and very efficiently produced the final typescript.

ACKNOWLEDGEMENTS

I have benefited greatly from the comments and suggestions of
Rod Coombs, Peter Halfpenny, Ian Howard, Christopher
Lawrence Freeman and many of the people who have
participated in the work. Many of the ideas discussed here have
been presented in seminars, and seminars given at the
Universities of Amsterdam, Aston, Bristol, Manchester,
Oxford, the Department of Social Economics at the 1981
Annual Meeting of the Danish Sociological Society and a
Third British Sociological Science Research annual conference on
Science Studies and I am grateful to the audience on these
occasions for their interest and suggestions. Much of the
empirical research upon the organisation of identical work in
different fields, well, benefited of one another during the
continually surrounding a different science-sentiment on
the workbench. I am more than happy also indebted to John
Allen, Billy Turner, Bob Card and Andrew McAleese who
cooperated to the appreciation of much a visual stimulant over
the months. This points to the practices of Christopher Freeman
and others, hoped that to Helen Walsh herself who has diligently
coped with the manifold tasks of the Section which it produced,
producing the final document.

CONTENTS

I

THE MODERN SCIENCES AS REPUTATIONAL WORK ORGANIZATIONS

INTRODUCTION

THE modern sciences have become increasingly important parts of the cultural systems of industrialized societies as producers and validators of knowledge claims about their natural and social environments. They dominate the construction of cognitive orders which serve as means of orientation and attempt to monopolize the production of true knowledge about the world. From being subsidiary to theology and philosophy as systems of understanding, the modern sciences have become the major knowledge-producing social institution in societies where scientific knowledge is highly regarded and its production strongly supported.[1] As such they have been the object of considerable discussion by philosophers, historians and, more recently, social scientists who have usually sought to understand their distinctive nature, patterns of development, and conditions of progress.[2] These discussions have traditionally assumed that scientific knowledge was essentially unitary, produced by distinctive methods and procedures, and justified in terms of a unique logic which guaranteed its truth value and progressive nature. Thus, they regarded the development of scientific knowledge as an epistemologically rational process which was not contingent upon the social conditions of its production and assessment. Intellectual change and development in the natural sciences and mathematics was excluded from the purview of the sociology of knowledge and the sociology of science which have tended to focus upon the social determinants of social thought and the social system which supported the generation of true knowledge of the natural world.[3] Most work in the sociology of science, in particular since the Second World War, has concentrated on the analysis of scientists' actions and patterns of their social organization in isolation from the knowledge they produced and changes in it.[4]

In the past decade and a half or so, the more celebratory and intellectualist approaches to the study of knowledge production and change in the modern sciences have become less dominant. Even some philosophers of science have begun to examine patterns of scientific discovery without reducing these to some logic of justification, although they still seem to be searching for *the* logical reconstruction of the discovery process.[5] The Anglo-Saxon history of science has started to treat the natural sciences less reverentially and more as a social phenomenon which can be analysed with the same historiographical procedures and techniques as other historical objects.[6] Similarly, the exclusion of intellectual change and development from the sociology of science has become less strict as European social scientists have resurrected the sociology of knowledge and sought to include the natural sciences in its programme.[7] At the same time, and no doubt not coincidentally, many national governments have begun to treat science as a national resource to be managed and directed for general political goals.[8] 'Science Policy' has become both an area of research and a set of administrative practices as the modern sciences have developed into a major, and expensive, social institution which requires 'steering' and monitoring by state agencies who are assisted by a variety of research groups and units.[9]

These changes in intellectual orientations, by no means universally accepted amongst philosophers, historians, and social scientists, have usually been associated with the impact of Thomas Kuhn's *The Structure of Scientific Revolutions* (first published in 1962 with a second enlarged edition in 1970) upon students of the modern sciences and upon the enlarged scientific labour force itself.[10] While some of Kuhn's points had been discussed earlier, notably by Polyani, Fleck and Bachelard,[11] his essay seemed particularly timely and appropriate in an era of organized, 'big' science. The large increase in national resources devoted to knowledge production and expansion of the higher education systems of many western nations heightened awareness of the domination of the natural sciences and the importance of knowledge production as an organized activity. No longer seen as definitely beneficent, science became the object of numerous analyses which used Kuhn's book to undermine existing intellectual élites and their understanding of scientific knowledge.[12] It is quite likely that these critical attempts to reorient the analysis of scientific knowledge would have occurred without his contribution, but they certainly relied heavily upon his approach in seeking to develop new understandings of how scientific knowledge has changed and 'progressed' which were not subservient to philosophical doctrines.

This was especially the case in the sociology of science in Europe and its revitalization of the sociology of knowledge. Kuhn's model of scientific change was extensively used to buttress the claims of the 'new' sociology of scientific knowledge and the sociological study of scientific communities which attempted to combine the analysis of social structures with changes in intellectual development. [13] A series of studies was produced in the early 1970s which focused upon the conditions supporting the emergence of new fields in the sciences and the circumstances which favoured state intervention in the direction of scientific research.[14] In particular, the 'Starnberg group' derived a three-phase model of knowledge development from Kuhn's book which suggested how governments could effectively influence the orientation of research in different fields.[15] In common with other attempts to use his model, this approach reproduced Kuhn's unitary view of the 'mature' sciences and his apparent belief that once a field became 'mature' through unspecified processes, it followed the same pattern of intellectual change regardless of its external circumstances.[16]

These two assumptions of uniformity and inevitability in knowledge development in mature fields prevented social scientists from studying variations between the sciences and investigating the social conditions in which radical intellectual change will occur. By implicitly or explicitly following Kuhn in presupposing the similarity of established sciences with modern physics and the inevitability of its pattern of intellectual change, many students of the sciences effectively removed the comparative analysis of scientific fields from their research agenda, together with any attempt at understanding the social circumstances in which revolutions were, or were not, likely to occur, in favour of focusing upon the processes by which new fields became established and produced correct scientific knowledge.[17] Although claiming their independence from epistemological domination and the validity of sociological analyses of intellectual changes, these studies effectively reproduced the privileged position of the natural sciences by accepting Kuhn's unitary model of knowledge development and its self-sufficient nature. It is difficult to see how a genuine sociology of scientific knowledge can be produced, though, without considering how intellectual production and assessment can change in different ways in different circumstances. Although a number of case studies of particular fields becoming established were produced in the 1970s, these were difficult to compare systematically and did not lead to a sociological account of how different sorts of knowledges were produced in different social contexts.[18]

The other major strand of post-Kuhnian[19] work in the European sociology of knowledge and science has focused more directly upon the issue of epistemological rationality and the implications of sociological reductionism in the analysis of intellectual change. In what has become styled as the 'strong programme' and the 'relativist programme'[20] of the sociology of knowledge, a considerable number of historical and contemporaneous studies have been undertaken to demonstrate the inadequacy of purely epistemological accounts of scientific judgements.[21] Essentially these have been concerned to show how scientists' judgements are inherently social so that intellectual changes are socially contingent rather than the outcome of epistemological rationalities and philosophical theories of scientific progress. Thus, scientists have been described as following their 'interests' and 'investments' in resolving controversies, and the production of scientific facts has been characterized as a highly socially contingent process of creating cognitive order out of disorder.[22]

However, the general implications of these studies are by no means clear and many authors seem content with showing the importance of non-epistemological factors in the resolutions of controversy and conflict. Few attempts have been made to draw comparisons between these case studies so that we could begin to analyse the social processes involved in generating and dealing with different kinds of scientific controversies in different situations, and it is not obvious in what respects scientific conflicts differ from other types of disputes. Equally, the recent studies of laboratory scientists negotiating the meanings of scientific facts offer little assistance for those seeking a comparative sociology of scientific change; indeed this goal is almost completely rejected in the Latour and Woolgar study where the construction of cognitive order is seen as an almost miraculous activity which defies further analysis.[23] The possibility that these laboratories may be unusual in their pattern of working, or historically variable in their organization, does not seem to have been taken seriously by these authors who make no attempt to situate their studies in the broader social context. The denial of the universality of homogenous, totalistic, and coherent scientific communities dominating the lives and thoughts of scientists, in the way that Kuhn and others suggest, need not imply the rejection of any sort of social structure beyond small groups and 'transepistemic arenas of research'[24] having some impact upon scientists' actions and judgements. The realization that some biomedical scientists organize their work in different ways from some physical scientists in the latter half of the twentieth century should surely have prompted consideration of why this is the case, what are its implications for knowledge production and organization, and how can we analyse such variations, rather than generalizing this particular point to all scientific fields and knowledge manufacture processes.[25]

Recent efforts at reorienting the study of scientific knowledge production and development, then, have led to a number of empirical studies of the emergence of new fields, of scientific controversies, and of the construction of scientific knowledge in particular circumstances as social phenomena without generating much comparative understanding of how different sorts of sciences become established and develop in different ways in different circumstances, or how the process of differentiation of knowledge production units itself occurred. I suggest that this sort of comparative understanding is an essential part of any adequate sociology of scientific knowledge which seeks to analyse how different knowledges are produced and changed; this book is intended to contribute to such an understanding by presenting an account of the modern sciences as particular kinds of work organizations which construct knowledges in different ways in different contexts.

In developing a comparative understanding of how different ways of conducting and validating research have become established and changed, some overall sociological view about the particular nature of the modern sciences as systems of generating and selecting intellectual innovations needs to be outlined and the major institutional developments, such as the growth of employment opportunities for knowledge producers, identified. This will be done in the rest of this chapter and the next one.

Within this broad characterization of the modern sciences as particular types of work organizations and control, there are, of course, substantial differences in the sort of research which has been, and is being, conducted in different fields, and in the way it is organized and controlled. These differences are sometimes seen as derivative from the subject matters of separate disciplines, such as the 'complexity' of the social world,[26] but are better construed, I suggest, as historically contingent variations which alter with changing circumstances and contexts. Thus, the evident lack of consensus on many key assumptions in the modern social sciences seems more comprehensible in terms of their close connections to everyday concerns and audiences, and a consequent plurality of standards and goals, than as a direct product of some essential properties of their subject matter.[27] Furthermore, the academic discipline as the basic unit of social organization of knowledge production is itself historically variable and is by no means an essential feature of the modern sciences. The developing domination of intellectual work by university professors during the nineteenth century had major consequences for the organization and control of research, as will be discussed in the next chapter, but was certainly not an inevitable concomitant of increasing scientific prestige and importance.[28] University-based disciplines are therefore only one type of knowledge production unit which unified reputational networks, employment structures, and training programmes in the late nineteenth and early twentieth centuries in many countries.

A broader and more general social unit of knowledge production and co-ordination is the intellectual field. These fields are conceived here as relatively well-bounded and distinct social organizations which control and direct the conduct of research on particular topics in different ways through the ability of their leaders to allocate rewards according to the merits of intellectual contributions.[29] While many intellectual fields did become entrenched in university departments and training programmes in the nineteenth century, not all did so,[30] and their identities are by no means always identical with employment or education unit boundaries. Such fields vary in the degree of cohesion and autonomy from other social structures, but constitute the major social entities which co-ordinate and orient research across a wide variety of situations and, increasingly, nation states. They reconstruct knowledge around distinct 'subjects' and their organization and change are crucial aspects of intellectual work and knowledge production in the modern, differentiated sciences. It is their patterns of organization, interconnections, and change which are the major concerns of this book.

In Chapters 3 and 4 I shall develop two sets of analytical dimensions for comparing intellectual fields, their internal organization, and the social contexts in which they are likely to become established. In Chapters 5 and 6 I shall combine these dimensions to identify seven major types of intellectual fields, their patterns of work organization and control, and the critical contextual factors associated with their development and institutionalization. Finally, in the last chapter I shall discuss certain interrelationships between intellectual fields, especially since they became linked to employment organizations such as universities, and how these have changed in the past hundred years or so.

Before continuing with the general characterization of the modern sciences as reputational systems of work organization and control, there are a few points which need clarification. First, the term 'science' is used here in a broad sense to refer to all forms of modern scholarship, rather than just to the natural sciences. This seems to me to be essential if we are to understand why these particular fields of research have become so dominant in our conception of knowledge and truth. Furthermore, the entrenchment of the natural sciences in the Prussian university system of the early nineteenth century followed the pattern established by the humanities and was marked by it. Late nineteeth-century and twentieth century ideals of 'pure' science and conceptions of knowledge production thus stem from general academic practices and structures and are not peculiar to, say, chemistry and physics.[31] Second, although the importance of industrial and state-directed sciences is considerable, and has had a major influence upon the development of research oriented primarily to truth ideals and controlled more through the formal communication system than by authority structures in employment organizations, I shall be focusing mainly upon this latter, 'public' science. This is because it is the public sciences which tend to dominate social conceptions of truth and knowledge. While legitimation and resources for the pursuit of truth are often sought in terms of utilitarian benefits by members of scientific establishments, they retain control over the interpretation and application of truth ideals, and it is their constructions and assessment criteria which legitimate many professionals' claims to expertise and knowledge. It is, then, the systems of knowledge production which are organized around the public communication of task outcomes and ideas which are the primary concern of this book. Thirdly, it is worth emphasizing that in talking of intellectual fields as the primary unit of organization concerned with the production and control of intellectual innovations, I do not mean to refer to Kuhnian notions of 'community' as developed by Hagstrom, Storer, and others.[32] Rather, fields are the social contexts in which scientists develop distinctive competences and research skills so that they make sense of their own actions in terms of these collective identities, goals and practices as mediated by leaders of employment organizations and other major social influences. Intellectual fields are here seen as the major forms of social organizations which structure the framework in which day-to-day decisions, actions, and interpretations are carried out by groups of scientists primarily oriented to public intellectual goals. They are variously structured and these variations are associated with differences in intellectual organization and occur in different circumstances. To understand how and why knowledges of different subjects vary and change it is essential to understand how and why the systems of their production and evaluation vary and change, and it is these which are the focus of attention in the comparative analysis of scientific fields.

SCIENTIFIC FIELDS AS SYSTEMS OF WORK ORGANIZATION AND CONTROL

The idea that scientific knowledge is produced by social actors in particular historical circumstances which constrain the nature of intellectual innovations considered to constitute knowledge is no longer especially novel. Ravetz, for example, has elaborated an account of scientific research as essentially a craft activity concerned with transforming intellectually constructed things and events with collectively organized methods and tools,[33] and other writers have described the social processes of construction of scientific 'facts' and manufacturing knowledge in particular areas of the modern sciences.[34] As such, scientific knowledge has become secularized and mundane in recent analyses of science: it is the product of an elaborate social organization which has become entrenched in educational institutions and employment markets rather than of isolated geniuses reading the book of nature. Thus, knowledge is produced and justified by particular methods which are acquired through lengthy training programmes and applied in particular locations called research laboratories and universities. It is the result of collectively organized work.

However, this realization of the socially produced nature of modern western scientific knowledge has rarely prompted detailed consideration of how such knowledge production has been differently organized and controlled in different circumstances. The Hagstrom exchange-recognition model, for instance, of the scientific community does not examine the production process of how information is used to construct new knowledge.[35] His account of competition focuses on recognition for past achievements rather than upon conflicts over the relevance and significance of research results for future work and so the direction of research. While the more recent accounts of scientific practices have concentrated on how new knowledge is constructed, and the social processes of resolving controversies, they focus mainly on local contingencies and conflicts without analysing the general system of work organization and control. Nor have they considered how different circumstances might result in different sorts of conflicts and patterns of fact construction.[36] Indeed, they rarely explore the historical peculiarity of the laboratory as a site for the generation of knowledge and seem to restrict scientific knowledge to the products of laboratory research, thus ignoring the field sciences and other areas.[37] Consequently, while scientific knowledge has become seen as the product of organized human activities in certain locations, it remains rather obscure quite how these activities are organized and controlled so that different patterns of work organization and control lead to different types of knowledge – for instance between the field sciences and the laboratory sciences.[38]

If we agree that scientific research is a type of craftwork which involves problem-solving on artificial objects,[39] then it seems reasonable to analyse the social organization which structures and controls this activity as a system of work organization and control which can be understood in a similar way to other forms of work organization. Thus, intellectual fields which constitute distinct spheres of knowledge production and evaluation are considered to be particular types of work organization sharing certain features with other types and understandable in comparable ways. Just as different patterns of work organization and control are in general associated with different environments, activities, and outputs,[40] so too we would expect differences between the sciences in the way they structure and co-ordinate research to be linked to variations in the organization of knowledge on the one hand, and to differences in their contexts and environments on the other hand. Equally, major changes in the contexts of scientific fields should have some impact upon how research is organized and controlled in them and so on the structure of intellectual innovations, just as large-scale changes in markets and communcation channels affected patterns of industrial organization and evolution.[41]

In considering the particular type of work organization and control which characterizes the modern sciences, a key feature is their commitment to producing novelty and innovations. Relative to other work organizations and systems of knowledge production, they institutionalize the dominant value of producing new knowledge which goes beyond, and is an improvement on, previous work. Rather than simply reinterpreting and elaborating past wisdom, modern western science is oriented to the construction of new and better intellectual artefacts which transcend earlier understandings. Thus intellectual obsolescence is built into the knowledge production system and old knowledge is devalued by new developments. This means that the outcomes of research tasks are inherently different and uncertain and the level of task uncertainty in the production system as a whole is greater than in most other work organizations. This in turn leads to a particular structure for organizing and controlling research which I term the reputational system.

An important aspect of this search for novelty and innovation is its extension to the methods of working on cognitive objects as well as to the outcomes of research. While many areas of work exhibit considerable uncertainty about the nature and meaning of task outcomes, the modern sciences are unusual in the degree to which they continually refine and alter their techniques and procedures so that practitioners frequently have to change their working practices. This constant innovation in how research is conducted obviously increases the overall level of task uncertainty and makes co-ordination and comparison of research results difficult, especially across research sites and national cultures.

This institutional commitment to novelty in the modern sciences is counterbalanced by their other major distinctive feature – the collective appropriation of task outcomes to produce new knowledge. Although much scientific knowledge is of course produced for consumption and use by employers, technologists, and the educated public, the most highly valued knowledge is produced for the consumption and use of colleagues in the process of producing innovations themselves.

Thus, scientists who seek the highest reputations as novelty producers have to convince powerful colleagues of their competence in following standard procedures and applying shared skills and of the significance and relevance of their work for collective goals. Intellectual innovations are valued in this system of knowledge production for their usefulness to researchers in producing more innovations. This means that task outcomes have to fit in with the aims and skills of others if they are to be highly regarded, and so the degree of novelty which can be produced in the modern sciences is restricted by the need to follow collective standards and be relevant to the work of colleagues. The critical point here is that research is valued to the extent that it affects, influences, and is essential for others' work to be successfully accomplished. Competent but insignificant research may be published but will not lead to positive and substantial reputations for intellectual achievements, and so the production of novelty is oriented to influencing and directing the work of colleagues in the sciences.

This emphasis upon the use of task outcomes to generate new knowledge distinguishes the sciences from many other areas of cultural production. Most cultural artefacts are generally judged by diffuse and broad standards of competence and importance which are shared among particular cultural élites but are only rarely evaluated according to their direct usefulness in creating new artefacts. Much literature, for instance, is based on a shared set of assumptions, linguistic conventions, and commonly recognized achievements which underlie reputations and assessments, but writers do not appropriate the products of other authors in creating their own nor is the renown of a writer derived from the number of famous colleagues who use his or her 'results'. So while cultural production systems in general are characterized by 'a constant and pervasive tension between innovation and control',[42] modern science is surely distinctive in the evaluation of intellectual innovations being so directly tied to their use in producing further novelties.

The institutionalization of this aspect of the modern sciences implies a considerable amount of intellectual autonomy and the ability of scientists to control sufficient resources to establish their standards as critical for access to rewards. As Randall Collins has pointed out,[43] this autonomy and control are fairly recent phenomena which enabled researchers to produce results primarily for one another and constitute the major audience for task outcomes. Critical to this growth in mutual dependence and exclusion of lay audiences and standards was the increasing control over the expanded secondary and tertiary educational systems exercised by intellectuals in many European countries during the nineteenth century. This enabled them to insist upon their goals and standards being crucial for training teachers and thus linked the control of jobs to intellectual achievements and renown. The general social prestige of scientific reputations was also increased by the role of universities in training and certifying the higher grades of state functionaries, especially in Prussia.[44]

As systems of work organization and control, the modern sciences are distinguished, then, by their commitment to producing novelty and innovations, on the one hand, and their co-ordination of research procedures and strategies through collective appropriation and use of their results, on the other hand. Novelty production is managed by allocating rewards on the basis of how successful a result is in influencing the direction and conduct of others' work so that scientists are constrained by the need to fit in with colleagues' plans and procedures. Only novelty is rewarded in this system of cultural production, but innovations have to be accepted and used by one's competitors for positive reputations and so strong collective control of intellectual change is maintained. The 'essential tension' between novelty and tradition, or co-operation and competition, is a notable feature of modern scientific work which results in a distinctive kind of work organization being characteristic of scientific fields. This reputational system of co-ordinating and controlling research is subset of the more general craft mode of work administration and shares some key features of professional systems of work control.

THE MODERN SCIENCES AS A TYPE OF THE CRAFT MODE OF WORK ADMINISTRATION

The organization and control of work in the sciences reflect general aspects of work structure and control in that issues of task formulation, differentiation, allocation, co-ordination, and evaluation are involved. Additionally, of course, authority structures and hierarchies are prominent features of work organizations in most industrialized societies and are no less observable in the sciences. The frequent implication in many studies of scientific communities that all contributions are assessed according to purely universalistic criteria which ignore the fame of their authors is negated by the many cases of differential treatment accorded to the ideas of established scientists compared to those of unknown contributors.[45] Accordingly we would expect many of the relations between task structures and control systems which hold for work organizations in general to apply to the sciences. For example, as systems producing novelty, the sciences are work organizations where the nature of the task and the meaning of task outcomes are highly uncertain and non-routine. Predictability and replicability of task outcomes are difficult and limited in extent. This feature of scientific work means that tasks can be differentiated and fragmented only to a restricted extent, and that they cannot be preplanned by managers outside the work-place to any great degree.

The high degree of uncertainty in scientific work clearly differentiates it from work in mass production industries and many other spheres. Control over work processes in the sciences is, on the whole, exerted by practitioners at the research site and is not governed by elaborate systems of rules and regulations established by external authority sources. Exactly what tasks are done, how they are done, and when they are done is decided by scientists rather than by any other group, and in particular not by administrative superiors in formal authority hierarchies. Work planning and execution in the public sciences are decentralized to individual workers who maintain considerable control over low-level goals and the use of particular procedures. The high level of task uncertainty renders fully bureaucratic systems of work planning and control inefficient and ineffective.

The non-bureaucratic nature of scientific work and its control can be clarified further with the notion of a 'craft system of work administration' as developed by Stinchcombe.[46] In certain important respects the sciences can be described as 'crafts' in distinction from bureaucratic systems of work planning and control and so subsumed under this general heading of work administration. Stinchcombe distinguishes between bureaucratic and craft systems of work administration largely on the basis of who controls the way work is carried out and whether it is planned in advance by non-direct production workers. He takes mass production industries and the construction industry as exemplary instances of the two systems. In the former, work is planned in detail in advance by administrative staffs while in the latter where, how, and when specific tasks are carried out is largely decided by the work crew on the basis of skills acquired during apprenticeships and certified by the craft unions. While bureaucratic systems of work administration require extensive clerical staffs and an elaborate, formalized communication system, in the craft system of work administration control over how the work is done is subcontracted to workers and foremen, and formal communications with the central administration are restricted to questions of product type and design and price. Economies of scale in the former case are achieved through standardization of tasks, but the construction industry relies on standardization of product type and of parts for its economies of scale.[47]

An additional difference concerns the authority system. In bureaucracies authority is, at least in principle, unitary. Both work goals and procedures are controlled by a single hierarchy in the employment organization. In contrast, authority in craft systems of administration is split between the employer and the certifying agency which produces and controls the skills required to do the work. Generally, employers control what the work is done for, and the disposition of the product, but they share authority over what skills are required and how they are applied and and co-ordinated with workers and the agency which certifies their skills. Because labour market status affects how work is done, and who does it, it becomes much more important in this case than in bureaucracies. The relatively small number of permanent posts in craft systems of administration compared to bureaucracies, combined with the influence of the extra-organizational status system on how work is done, mean that for many workers their permanent labour market status is more important than their temporary employment status.

Stinchcombe suggests that these two forms of administration are two types of 'rational' work administration which are equally valid as efficient forms of work control given different circumstances. While the considerable overhead costs incurred in the bureaucratic form are justifiable where production runs are long, markets are large, and work-flow is predictable and relatively stable, they become too burdensome where these conditions do not hold. Instability in work and income-flow coupled with considerable variability in product and volume mix and work-force composition render the full panoply of Taylorian planning and control systems inappropriate and the subcontracting, craft system more 'rational'. Here, the uncertainty and variability of the work-flow is managed through short-term employment statuses and decentralizing considerable control over work processes to the direct production workers, and indirectly to the skill-certifying agency.

In many respects, such as work-flow instability and variability, scientific work is similar to craft work in the construction and similar industries. Equally, control over work procedures tends to be the preserve of practitioners rather than being governed by formal rules established by an administrative hierarchy. Additionally, external status tends to be more important than immediate employment status for many scientists, and, indeed, often determines it in the public sciences. On these grounds, then, the sciences seem closer to the craft form of work administration than to the bureaucratic form. However, there are a number of obvious differences which highlight some of the peculiarities of scientific work.

First, the existence of a labour market for scientific skills is a relatively recent phenomenon. Continuous payment for doing original research, and provision of facilities, were not widespread before the middle of the nineteenth century when university positions for researchers become established on a fairly wide scale. Most posts held by scientists were intended for a variety of purposes, usually focused on administrative and military requirements, or were pensions awarded for earlier work rather than salaries paid while research was being conducted.[48] Some of the Royal Academies awarded prizes for research, which could be seen as a form of labour subcontracting, but these were inadequate for personal support and research apparatus provision on a long-term basis.[49] Furthermore, when payment did become a regular feature for the conduct of research it was usually tied to permanent empoyment positions rather than to short-term employment contacts based on specific projects. While this was due to a number of reasons, a major factor was the high degree of task uncertainty in scientific work which differentiates it from many crafts.

This second difference is not simply a matter of degree. The institutional imperative in science of producing new, original knowledge means that the skills inculcated during training programmes do not remain fixed and stable over working lifetimes but are being constantly developed and altered. The instability and variability of the work-flow in the sciences does not simply arise from product and market complexities but stems from the commitment to innovation. Every new result and task outcome in science must be different from preceding ones if it is to be regarded as a contribution to knowledge. Variation is thus built in to the activity itself and the skills used in scientific work must be capable of generating a succession of new and different results if they are to make useful contributions. They are, therefore, more general and wide-ranging, and lead to a greater variety of outcomes than do most craft skills. Employment of such skills will not result in definite, clearly bounded outcomes but rather to the production of particular types of knowledge which can be variously interpreted and understood. The predictability of results obtained from using certain skills and materials is therefore less in the sciences than in most crafts. The nature of the product is difficult to specify clearly in advance and is subject to negotiation when it has emerged. Subcontracting scientific work on a project basis where the employer specifies the product beforehand is not very straightforward in these circumstances. Instead, employers have preferred to grant permanent employment status to scientists and rely on a combination of bureaucratic controls and professional socialization. Goals are set by the administrative hierarchy in much industrial research but work processes are, usually, decided by scientists on the basis of their training. Adherence to organizational objectives is ensured by structuring job statuses into a hierarchy which often leads into managerial positions, thus reducing the importance of labour market status in favour of organizational status. Supervision of scientific work is exercised by higher-rank scientists at the work-place, and although work is not systematically preplanned by administrators it is often quite narrowly specific and discretion limited. Much scientific work in industrial laboratories, then, resembles the craft mode of administration, but bureaucratic systems are also in evidence and are sometimes directly applied as in the case of formal planning models.[50]

In the public system of science direct supervision of research also occurs at the work-place, often in a hierarchical form, but it is supplemented by an elaborate system of formal communication and publications. This system is the third major difference between the sciences and the craft system of work administration. Control over work goals and procedures in the public sciences is exercised more through this formal system than through employment hierarchies – although empirically they are often difficult to separate. This does not make science bureaucratic in the sense discussed above because this formal system does not plan how work is to be carried out and is not under the control of non-direct production workers. Additionally, it is not identical to the employment structure as a control system and, indeed, often functions in contradictory ways to it. Furthermore, although frequently controlled by academics who also control the training and certifying systems, it need not always be identical with these systems and recently has shown signs of developing some autonomy from them.

The importance of this public communication system in the sciences highlights the innovatory nature of research and the uncertain nature of task outcomes. In effect, the constant changes in work procedures and purposes are managed by a very high degree of decentralization of control over work processes to the individual scientist which makes the organization very flexible and responsive to variations in the environment, coupled with a formal reporting system which enables task outcomes to be compared and co-ordinated. This reporting system relies on standardized symbol structures to reduce ambiguity and facilitate comparisons. Control is exercised collectively through this sytem by imposing standards and criteria on the evaluation of results, and hence directing work along certain lines to the exclusion of others. The perpetual generation of uncertainty in the sciences is monitored and controlled through this formal system which ensures that research never becomes too 'original'. To be recognized as contributions to knowledge, in a society where science monopolizes the production and validation of knowledges, research results must follow current priorities and use accepted work procedures to be accepted for publication. Innovation and novelty is thus always tempered by the exigencies of the control system.

THE MODERN SCIENCES AS A TYPE OF PROFESSIONAL WORK ORGANIZATION

The relatively high level of task uncertainty in the sciences and the corresponding degree of personal autonomy of scientists in employment structures have led many observers to see them as professions. However, the meaning of this term is unclear in many accounts and there is a large literature on the different defining attributes of professional occupations.[51] To clarify further some of the key characteristics of the sciences as forms of work organization, I shall now briefly discuss the similarities and differences between the sciences and professional work organisations.

Generally, it seems reasonable to locate professions as a subset of the craft system of work administration. Particular skills are inculcated by exclusive institutions and certified by professional agencies. These skills monopolize particular activities which are accorded high social status and corresponding high rewards. Work in the professions is controlled by practitioners in accordance with their training, but the purpose to which professional skills are put is controlled, to varying degrees, by lay clients and employers. How the work is carried out is the preserve of the worker, as in craft systems of work administration, but the selection and co-ordination of skills is often controlled by non-professionals. In some of the consultancy professions, such as private medicine and architecture, the certifying body exerts considerable influence on the conditions and terms of employment so that professionals are able to define the nature of the service and substantially control the evaluation to task performance, but this is a matter of degree and changes over time. Larson, for example, emphasizes the crucial importance of prospective professions constituting and controlling a market for their expertise[52] and yet notes the expansion of permanent employment status for many professions and their subjection to formal authority hierarchies. Labour market status as, say, an accountant may be important for access to particular posts in large organizations and may guarantee a considerable degree of discretion over how work is conducted but the allocation and co-ordination of tasks is often controlled by employers. Professional status here ensures that work processes are controlled by practitioners but goals and performance are decided by employers. Professional work organizations, then, are systems of craft administration of work where the certifying agency has some control over the conditions and terms of employment and is able to influence the evaluation of task performance. Authority over how skills are to be used, co-ordinated, and evaluated is shared between the professional groups and clients or employers. How this division takes place varies between professions and is subject to frequent negotiation and conflict between collegiate bodies and buyers of professional skills.

Science, according to Collins,[53] is a strong profession. Here, self-conscious and self-regulating colleague groups controlling intellectual innovations are based on their power to validate the expertise, and thus mediate the careers, of members. Material rewards are dependent on professional status and reputations so that workers have to follow the dictates and priorities of their colleagues. In science, practitioners' careers depend on convincing others that they have made substantial contributions to the knowledge goals of the field and so the collegiate group exercises strong control over what work is done and how it is done. Employers and clients in these situations allow autonomy to workers because of the high degree of task uncertainty and because of the ability of professional skills to reduce uncertainty to some extent. In general, the more unpredictable are task outcomes, the greater particular professional skills can reduce uncertainty compared with other skills, and the more important and valuable is the reduction of uncertainty to clients and employers, the more autonomy over work goals and procedures do professional groups have and the greater is their power over practitioners.

Just as professions vary in their autonomy and power over members, so, too, do scientific fields differ in the extent to which these conditions are met. Additionally, of course, different historical conditions affect the ability of professional groups to form and dominate work processes. As I have already mentioned, payment for scientific work on a continuing basis was unusual before the nineteenth century and so the ability of scientists to control access to material rewards through intellectual reputations was rather limited. Scientific careers as a sequence of reputational statuses had relatively little impact on personal income over a working lifetime until chemical consultancy became established as a means of earning a living in the early nineteenth century,[54] and the power of the collegiate group was correspondingly restricted. Partly because of this, the sort of knowledge that was produced before jobs became tied to scientific reputations, i.e. professionalized,[55] tended to be less oriented to collegiate concerns and priorities. It seems not unreasonable to suggest that the degree of innovation and novelty of knowledge claims was higher then than it has since become, just as Berman has discerned a difference between the sort of science produced by 'amateur' scientists in the nineteenth century, mostly in England, and that emanating from 'professionals'.[56] As strong professions, then, the sciences have varied in strength of control over work and careers.

If we restrict our attention to those periods in which high scientific reputations have led to material rewards then the sciences – or the 'mature' ones as Collins suggests[57] – do seem to function as strong professions in terms of collegiate control over work processes, objectives, and the evaluation of task performance. However, there is one important difference between the sciences and what are usually thought of as the major professions. In the professions the production of skills occurs at the beginning of the professional career, and the nature of the skill is regarded as relatively constant and fixed throughout the individual career. While scientific skills are also inculcated at the beginning of the working life, they undergo considerable change during a single career so that scientists constantly have to 'keep up' with the work of others and acquire new methods and techniques to a much greater extent than do other professional groups. What might have been regarded as a contribution to knowledge when a scientist began his or her career is rejected as unoriginal or irrelevant at a later stage. Collegiate groups in the sciences, then, do not simply produce skills and rely on normative means of work control to ensure they are properly applied, they also have to monitor research results and co-ordinate task outcomes in a systematic way. While other professional groups control work through producing particular skills which remain stable and so, to a large extent, can be taken for granted, and focus on the terms and conditions governing their employment and use, the sciences generate instability in work processes and so have to organize a system of feedback and control over work to ensure coherence and co-ordination. Without the formal communication and control systems, the pursuit of originality could fragment scientific fields.

The public reporting system in the modern sciences heightens collegiate control over work processes. While all professions, and indeed most crafts, control the conditions and terms of employment, and some control the evaluation of task performance and the purposes of employment, none controls everyday performance of tasks and monitors task outcomes in the way the sciences do. The extensive reliance on self-reporting in science should not obscure the powerful effects of the publication system on the conduct of research. To convince colleagues of the importance of one's work it first has to be published, thus ensuring conformity with public norms and criteria, and, second, it has to be used by them in their own research. The more important it is thought to be, the more competitors will both try to develop and discredit it. Competitive pressures ensures that new results and ideas which seem important will be used, transformed, and manipulated. If expectations are not fulfilled during these processes, the results will be rejected and ignored. This is especially likely with task outcomes which counter currently accepted views. While few results actually are checked to this extent, and most are ignored anyway, the emphasis of the system on fame and fortune following from convincing a large number of influential colleagues of the importance of one's work ensures general adherence to current procedural norms. If work is to make an impact it must be more than simply routine which will be published and then ignored. It must influence and direct the work of others. Given that they have a vested interest in down-grading others' work to the exaltation of their own, such influence is likely to be acknowledged only after substantial evaluation and testing. Faking results may produce mediocre reputations by enabling publication but is unlikely to lead to glory. The frequently attested obsession with checking and counterchecking one's results before allowing them to be published, and extensive pre-publication secrecy in many fields, are manifestations of this control system.[58]

This leads to a further difference between the modern sciences and most professional work organizations. Although practitioners compete with one another for market advantage, either directly for clients or through the professional reputational system, they do not publicly compete for influence and control of others' work by publishing the results of their own work. While the scope and degree of competition varies between the modern sciences, the competitive pursuit of prestige and influence among a particular audience seems an inescapable characteristic of them. Generally, the acquisition of professional skills is assumed to guarantee competence in most professions so that practitioners do not need to demonstrate their importance by continually communicating their successes to their colleagues. In science, though workers have to show their ability to control important areas of uncertainty in approved ways if they are to remain influential.

This continual dependence on colleagues for approval and recognition throughout one's research career represents a much higher degree of work control by professional agencies than that encountered in most professions. How, when, and where tasks are carried out may still be very much under the control of the individual scientist, but the results have to conform to public criteria of acceptability, and be accepted by colleagues, if continued participation in professional work is to allowed. The degree of individual autonomy from the professional group is thus much lower in these 'mature' sciences than in other professional fields. Evaluation of successful task performance is not left to the individual practitioner but must be negotiated with other professionals to an extent which is considerably greater than in other areas of professional work. In this sense the community and mutual orientation of the sciences is stronger than elsewhere; scientists are engaged in continual debates and conflicts among themselves about their work in a way which seems largely absent among other professional groups. This strong community structure, however, is much more competitive and conflictual than appears in some accounts and these conflicts have substantial consequences for what emerges as scientific knowledge.[59]

In summary, then, science is a professional work organization in so far as it controls how work is carried out, how it is evaluated and its criteria and procedures govern access to material rewards. Clients and employers may buy scientific skills for their own purposes but have to rely on scientists' judgements in recruiting personnel, allocating rewards, and assessing outputs. It differs from other professions by constantly monitoring task outcomes through an elaborate formal communication system which controls work to a greater extent than elsewhere, and by institutionalizing a high degree of mutual competition for influence and significance based on the importance of task outcomes. Practitioners in the modern 'mature' sciences are much more oriented to their colleagues, and their opinions, in their work than are most professional practitioners. These characteristics are strongly connected to the relatively high degree of task uncertainty and rapid change in skills which arise from the commitment to novelty and innovation.

THE MODERN SCIENCES AS SYSTEMS OF WORK ORGANIZED AND CONTROLLED THROUGH REPUTATIONS

The modern public sciences, then, share, many characteristics with the professions and craft systems of work administration but also differ in some important respects. These differences suggest that they constitute a distinct type of work organization and control in which research is oriented to collective goals and purposes through the pursuit of public scientific reputations among a group of colleague-competitors. In such reputational work organizations, the need to acquire positive reputations from a particular group of practitioners is the main means of controlling what tasks are carried out, how they are carried out, and how performance is evaluated. Here, work is conducted with a view to convincing fellow researchers of the importance and significance of the results and hence enhancing one's own reputations. Jobs and resources are allocated largely according to reputations in the organization so that status in employment organizations is dependent on one's reputation in the wider 'community'. Rather than careers being patterns of succession in organizational statuses organized in an authority hierarchy, they are constituted by a succession of reputations in one or more intellectual fields.

Control over how work is done, and for what purposes, through reputations implies more than collegiate interdependence for recognition. The Hagstrom model of exchange in science seems to suggest that scientists 'politely exchange gifts of information' or recognition, as Collins put it.[60] This model of colleagues of equivalent status and resources engaging in mutual back-scratching ignores the stratification of authority in the sciences and unequal distribution of resources. It also disregards the effect of differential recognition and rewards on the system of production of scientific knowledge and thus 'forgets' the crucial point about reward systems, that they are instruments of control. The search for reputations in the sciences, then, just as in the arts or other systems of cultural production, is not simply for mutual pats on the back but for power over knowledge goals and procedures. Reputations are won by persuading the relevant audience of the importance of one's work and so affecting their own priorities and procedures. Having a high reputation implies an ability to have your own views and ideas accepted as important so that others follow your direction. It also implies an ability to affect the allocation of research resources and, indirectly, jobs in work organizations where reputations control facilities. Struggles for reputations, then, involve battles over resources and priorities. Equally, rather than simply offering research results upon some neutral and impervious market for reputations, scientists engage in various strategies, with varying amounts and sorts of resources, to manipulate actively others' opinions and evaluations. Several recent studies of negotiations among scientists have emphasized this aspect of modern scientific work.[61] Reputational work organizations, then, are characterized by incessant attempts at getting attention and imposing ideas and concepts on colleagues. The particular ways in which these struggles are organized in different fields result in different patterns of intellectual organization and change. Where, for example, researchers have a wide variety of legitimate audiences for their work, including educated laymen, and research skills are not highly standardized, as in many of the human sciences, the need to co-ordinate research results with those of a particular group of colleagues to gain positive reputations is limited, and so contributions to intellectual goals are relatively diffuse and divergent. Integration of task outcomes around common objectives is, therefore, not likely to be very high in such fields.

In the modern sciences research objectives and procedures are controlled by insisting that only contributions which have been published in collegiate journals constitute scientific knowledge. Work which fails to match expectations institutionalized in the public system will not appear in print and hence cannot lead to high reputations and influence. Where such reputations do mediate access to material rewards, especially the case in late nineteenth and twentieth century industrialized societies, research becomes strongly oriented to collective goals and considerable competition is likely. Reputational control through elaborate formal communications systems can be quite high in these circumstances, as in much of modern chemistry and physics. A high degree of practitioner autonomy over exactly how tasks are carried out, i.e. control over work processes in Stinchcombe's approach, is combined with strong collegiate control over the purposes of work and the evaluation of task outcomes in these reputational work organizations. The high degree of task uncertainty which is generated by the pursuit of originality in science is here tempered by the need for collective certification of results and co-ordination of research with that of others to gain reputations. The formal communication system and the importance of scientific reputations ensure that work in these organizations is oriented to collective goals and is reported in a form which enables co-ordination of task outcomes from a wide variety of work-places.

The extent of originality and novelty in research goals and procedures is thus restricted by the need to convince specialist colleagues of the significance of one's work in reputational work organizations. While task uncertainty may be higher in the modern sciences than in other spheres of professional work, it is none the less constrained by strong collegiate control of reputations and the communication system. Contributions which do not match the priorities and interests of powerful groups are unlikely to be published or used and so wide-ranging systems of thought which call into question the beliefs, commitments, and techniques of colleagues will be only rarely produced in fields where skills are highly standardized and background assumptions widely shared. If scientists cannot obtain jobs or promotion without gaining high reputations from particular audiences, and those audiences have systematic criteria for the evaluation of contributions, few are going to risk their future by trying to publish material which deviates widely from current orthodoxies.

A major manifestation of the way reputational control limits the originality of contributions to collective intellectual goals is the necessity of referring to the previous work of colleagues. While this may be necessary to avoid prolix redundancies in the text, it is also a way of exerting social control over novel ideas. By insisting that authors refer to particular scientists and currently established evidence, reputational organizations ensure that work is not too far removed from the aims and procedures of the dominant group. The degree of innovation is thus diminished and constrained by the necessity of showing how new contributions fit in with, and are relevant to, existing knowledge. In a sense citations are a way of ritualistically affirming group goals and norms, of demonstrating group membership and identity. Rejecting the 'obvious' and 'relevant' precursors means rejecting group goals and disociating oneself from current conceptions of what is known. Such rejection is more likely to happen in loosely structured fields where alternative audiences and reputational organizations are available, as in many of the human sciences where lay audiences sometimes affect the standards according to which the significance of intellectual products are judged.

The development of strong reputational organizations controlling access to jobs and research facilities in the nineteenth century encouraged specialization of topics and skills in the sciences.[62] The norm of originality obviously leads to work differentiation as practitioners seek to demonstrate their novelty and innovation but this differentiation can take a number of forms, including outright rejection of previous work and approaches. In fields where scientists compete for reputations from one particular dominant group, though, the range of intellectual novelty which will develop is likely to be quite limited, for the reasons already stated. Innovations in such fields tend to be restricted to one dimension of cognitive variation, and often one which has few major implications for the work of others. High dependence on one group of colleagues for reputations, and hence material rewards, encourages many scientists to limit their originality to minor variations of procedure and topic or material so as to minimize the risk of rejection and conflicting with powerful colleagues. Specialization of techniques of systems of study is one way producing original work without threatening established interests and commitments. Of course such a strategy also reduces the likelihood the major reputations will be acquired but this may be unrealistic for most researchers anyway in highly stratified and monolithic fields such as twentieth-century physics. In these sorts of fields, often dominated by a relatively small number of departments and élite locations, radical contributions from the periphery are not often likely to result in high reputations and so those making up the bulk of the scientific labour force may well resign themselves to producing minor variations on established themes. Elite scientists may, of course, adopt a more innovative strategy with greater chance of success but even they will be constrained by the collegiate orthodoxy entrenched in journals, textbooks, and training programmes.[63] Establishing new sub-fields will be easier than attempting radically to alter dominant perspectives, and so intellectual change in these fields is likely to take the form of differentiation and specialization rather than revolutionary overthrows of established doctrines.[64] Research becomes more specific and restricted in scope, then, when scientists compete for reputations in highly co-ordinated and centralized fields.

CONDITIONS FOR THE DEVELOPMENT AND ESTABLISHMENT OF THE MODERN SCIENCES AS REPUTATIONAL WORK ORGANIZATIONS

The development of a strong degree of reputational control over work goals and task outcomes in intellectual fields depends on a number of factors, such as the existence of a common communication system, control over resources, and exclusive rights of assessment of results. These in turn depend on science as a whole having gained sufficient autonomy from other groups of cultural producers to impose its criteria on task outcomes and some control over the allocation of resources. Where scientific establishments have to share control over cognitive orientations in a society with other groups, and where they cannot award substantial resources to practitioners through the reputational system, they will not be able to control research priorities and procedures to any great extent. Thus the conditions for establishing scientific fields as distinct systems of work organization and control can be summarized under three main headings: *(a)* scientific reputations need to be socially prestigious and to control access to critical rewards, *(b)* each fields has to be able to set particular standards of research competence and craft skills and *(c)* each field has to control a separate communication system.

The formation of distinct reputational communities within science depends on both scientific reputations becoming socially prestigious and on researchers who are working within the current accepted definition of science being able to control reputations through the assessment of specific contributions by particular, rather than science-wide, criteria. This is partly a matter of sheer numbers in that the more practitioners there are, and the more they are oriented to collective opinion, the more competition for reputations will lead to specialization because the likelihood of convincing a large number of colleagues of the importance of one's world view is less than that of convincing a small group of the importance of one's contribution to a slightly narrower field. It is also a matter of reputational groups being able to organize themselves as distinct collectivities with their own communication system and evaluation criteria, and to claim the legitimacy of their work as science. The more important science became in modern societies, and the more it tended to monopolize claims to the production and validation of truth claims, the more crucial this last condition became.

Reputations in the special sciences are valued, then, when science as a whole has some social prestige and authority and when the particular field can award reputations which are widely seen as scientific. If an area of research is regarded, both within science and in the cultural system in general, as largely peripheral to doninant conceptions of science, then reputations in it will be correspondingly devalued and practitioners will seek additional audience elsewhere, either in other fields or among other groups. For an intellectual field to function as a distinct reputational organization then, it must be able to claim high status for its reputations. This means that when science has become prestigious, fields compete for control over the central ideals of science so that they can exemplify them and so dominate the award of scientific reputations. It also results in embryonic science, such as many human sciences in the late nineteenth and early twentieth centuries, imitating many of the features of established science in order to get their reputations accepted and highly valued. To control practitioners leaders of these fields have to be able to offer the accolade of being scientific, and so sometimes encourage ritualistic reproduction of work practices which are commonly considered to be scientific such as statistical analysis and mathematical manipulation. The more important it becomes to cultural producers to be thought scientific, the more such imitative behaviour seems likely and the less tolerance there will be for deviant notions of scientificity and large-scale criticisms of current scientific ideals.[65]

As well as awarding reputations which are highly valued within science, and, through the general prestige of science, throughout society, intellectual fields must have distinctive work procedures if they are to function as reputational work organizations. In a similar way to other professional work organizations, the sciences need to control particular craft methods of work which differentiate them from other concerns and enable them to control access to reputations in particular fields. Examples of these methods are: textual exegesis in philology, participant-observer fieldwork techiques in social anthropology, analytical tools in chemistry, and *drosophila* breeding in genetics.[66] Such skills and techniques are exclusion devices. These work procedures also need to be standardized to some minimal extent because task outcomes have to comparable across work-places and, to some extent, predictable. Without some standardization of techniques and methods scientific fields could not co-ordinate and control task outcomes and so construct systems of knowledge. They also could not award reputations because there would be no way of assessing the relative merits of different contributions since each would be wholly idiosyncratic. While innovation in technical procedures is endemic in the sciences, then, its degree is constricted by the need to ensure comparability and acceptance of results of colleagues. So each field has to have some particular skills which exclude outsiders and enable results to be compared and evaluated in terms of their significance for collective goals.

The third necessity for intellectual fields to become established as distinct reputational work organizations is a standardized symbol system for reporting task outcomes. These systems perform two functions. First, they differentiate work in one field from that in others and so act as exclusion devices. Second, they enable result to be communicated across research sites relatively unambiguously and so ensure that practitioners can co-ordinate their research strategies without extensive personal contact. For reputations to be established and transmitted where task uncertainty is fairly high, results need to be expressed in a highly structured language which reduce ambiguity and enables their implications for other work to be discerned relatively quickly. The more formalized are such languages, and the more restricted they are to particular scientific fields, the more control reputational organizations have over their members since alternative audiences are less likely to consider results expressed in esoteric symbols, and the linguistic skills necessary for making competent contributions in a particular field will not be usable elsewhere.

In summary, reputational work organizations exert strong control over their members when they have a relatively high degree of autonomy from the dominant culture, and from other professional groups, when they control the acquisition of standardized skills which are necessary to produce acceptable task outcomes and when they control a standardized symbol system which monopolizes the communication of results and the means of obtaining reputations. Finally, they must mediate access to valued cultural and material rewards if they are to maintain the allegiance of practitioners.

SUMMARY

The points made in this chapter can be summarized in the following way:

1. Scientific knowledge has increasingly come to be seen as the product of the social transformation of intellectually constructed objects and scientific change is increasingly viewed as the outcome of social processes of negotiation, conflict, and competition. However, the considerable number of empirical studies of scientists' actions and beliefs conducted over the past fifteen years or so rarely involve a comparison of different fields or historical periods. Furthermore, the general framework which struc-

tures scientists' activities and perceptions has received little attention.

2. Viewing scientific research as a form of work implies its susceptibility to comparative analysis as a particular type of work organization and control which structures the production and evaluation of knowledge claims in different ways in different circumstances. So differences and changes in scientific knowledges can be understood in terms of differences and changes in the system of their production and evalution considered as types of work organization termed intellectual fields. Fields organized and controlled in different ways produce differently organized knowledge and become established in different contextual circumstances.

3. As a particular system of work organization and control, modern science is distinguished by its combination of continual novelty production – and hence high task uncertainty – with strong collective co-ordination of task outcomes through access to rewards being controlled by reputations based on the utility of results for colleagues' research.

4. The high level of task uncertainty in scientific research extends to work procedures in addition to task outcomes. It is therefore greater than in many craft systems of work administration and professions. While sharing many features of these types of work organization, such as practitioner control of work processes and professional certification of skills, the modern sciences differ in their continual revision of research practices and methods. Thus, the initial acquisition of research skills in training programmes has to be supplemented by additional improvements and modifications throughout a researchers' career if she/he is to continue to make competent contributions to collective intellectual goals.

5. This endless revision of work procedures, and the need to convince specialist colleagues of the relevance and importance of one's results for their concerns in order to gain a positive reputation, result in co-ordination of task outcomes being achieved through a formal public communication system. This system both connects research results from

different production sites concerned with common problems and provides the arena for conflicts over reputations and interpretations. It is the major agency of social control of competence standards and work process as well as being the locus of negotiations over intellectual goals and priorities.

6. As reputational systems of work organizations and control, the modern sciences co-ordinate and direct research through the allocation of rewards according to the competence and significance of public contributions to collective intellectual goals as determined by their utility in producing new contributions. Intellectual reputations in each fields mediate access to material rewards and depend on the importance of an individual's work for his or her specialist colleagues' research. Sciences vary in their degree of reputational control of research and in the way it is organized. These differences are related to variations in the contexts of scientific fields.

7. Scientific fields become established as distinct reputational organizations when:

   (a) science as a whole is socially prestigious and scientific reputations lead to rewards,
   (b) the particular field is able to control access to rewards through its reputations,
   (c) the particular fields is able to control competence and performance standards and has demonstrably distinct research skills which reduced some uncertainty, and
   (d) the particular field has a distinctive language for describing cognitive objects and communicating task outcomes which reduces lay participation in the assessment of contributions and enables results from different production sites to be compared and co-ordinated.

Notes and references

[1] Not, of course, without conflicts which affected the nature of scientific knowledge which become established. See, for example, W. v. d. Daele, 'The Social Construction of Science', in E. Mendelsohn *et al.* (eds), *The Social Production of Scientific Knowledge*, Sociology of the Sciences Yearbook I, Dordrecht: Reidel, 1977.

[2] On the United States' history of science, see A. Thackray, 'The Pre-History of an Academic Discipline: the Study of the History of Science in the United States, 1891–1941', *Minerva*, XVIII (1980), 448–73; on the early philosophy of science, see L. Laudan, 'Peirce and the Trivialisation of the Self-Correcting Thesis', in R. N. Giere and R. S. Westfall (eds.), *Foundations of Scientific Method: the Nineteenth Century*, Indiana University Press, 1973; D. L. Hull, 'Charles Darwin and Nineteenth Century Philosophies of Science', in *idem;* L. Laudan, 'The Sources of Modern Methodology', in R. E. Butts and J. Hintikka (eds.), *Historical and Philosophical Dimensions of Logic, Methodology and Philosophy of Science*, Dordrecht: Reidel, 1977. On the early sociology of science, see S. B. Barnes and R. G. A. Dolby, 'The Scientific Ethos: a deviant view point', *European Journal of Sociology*, XI (1970), 3–25; M. King, 'Reason, Tradition and the Progressiveness of Science', *History and Theory*, 10 (1971), 3–32; R. D. Whitley, 'Black Boxism and the Sociology of Science', in P. Halmos (ed.), *The Sociology of Science*, Sociological Review Monographs 18, Keele University Press, 1972.

[3] The classic formulation of the sociology of knowledge is, of course, by Karl Mannheim, *Ideology and Utopia*, London: Routledge & Kegan Paul, 1960, (1936), ch. 5. See also his *Essays on the Sociology of Knowledge*, London: Routledge & Kegan Paul, 1952, chs. 4 and 5.

[4] As discussed by R. Whitley, op.cit., 1972, note 2.

[5] See, for example, the two volumes of papers edited by T. Nickles, *Scientific Discovery*, Dordrecht: Reidel, 1980.

[6] On recent history of science in the United States see: N. Reingold, 'Clio as Physicist and Machinist', *Reviews in American History*, Dec. 1982, 264–80. On the use of prosopographical techniques in the history of science see L. Pyenson, "Who the Guys Were' Prosopography in the History of Science', *History of Science*, XV (1977), 155–88; S. Shapin and A. Thackray, 'Prosopography as a research tool in History of Science', *History of Science*, XII (1974), 1–28.

[7] On recent developments in the sociology of science, see M. Mulkay, *Science and the Sociology of Knowledge*, London: Allen & Unwin, 1979.

[8] M. Mulkay suggests that the growing concern of various states with the direction and management of science was connected to the growth of 'science studies' and similar areas of research in his 'The Sociology of Science in the West', *Current Sociology*, 28 (1980), 133–84.

[9] For a recent summary of this work see S. S. Blume, *Science Policy Research*, Stockholm: Swedish Council for Planning and Coordination of Research, 1981.

[10] T. S. Kuhn, *The Structure of Scientific Revolutions*, Chicago University Press, 1962.

[11] As briefly mentioned by, *inter alia*, H. Martins, 'The Kuhnian 'Revolution' and its implications for sociology', in T. J. Nossiter *et al.* (eds.), *Imagination and Precision in the Social Sciences*, London: Faber & Faber, 1972.

Martins's essay is one of the few systematic discussions of Kuhn's book from a sociological point of view as distinct from the numerous philosophically inspired analyses. See G. Bachelard, *La Formation de L' Esprit Scientifique*, Paris: Vrin, 1972 (1938); *Le Nouvel Esprit Scientifique*, Paris: P. U. F. 1976 (1934); L. Fleck, *Genesis and Development of a Scientific Fact*, Chicago University Press, 1979 (Schwabe, 1935); S. W. Graukroger, 'Bachelard and the Problem of Epistemological Analysis', *Stud. Hist. Phil. Sci.*, 7(1976), 189–244; M. Polyani, *Personal Knowledge*, New York: Harper & Row, 1964 (Chicago, 1958); *Science, Faith and Society*, University of Chicago Press, 1964 (O.U.P., 1946).

[12] For an account of this process in United States' anthropology, see B. Scholte, 'Cultural Anthropology and the Paradigm-Concept: a brief history of their recent convergence', in L. Graham *et al.* eds), *Functions and Uses of Disciplinary Histories*, Sociology of the Sciences Yearbook 7, Dordrecht: Reidel, 1983.

[13] See, for example, many of the papers in R. Whitley (ed.), *Social Processes of Scientific Development*, London: Routledge & Kegan Paul, 1974.

[14] See, for instance, the case studies in G. Lemaine *et al.* (eds.), *Perspectives on the Emergence of Scientific Disciplines*, Paris: Mouton, 1976.

[15] See G. Böhme *et al.*, 'Finalization in Science', *Social Science Information*, 15 (1976), 307–30. On the subsequent extended and politicized debate over this approach see F. Pfetsch, 'The 'Finalization' Debate in Germany', *Social Studies of Science*, 9 (1979), 115–24 and A. Rip, 'A Cognitive Approach to Science Policy', *Research Policy*, 10 (1981), 294–311.

[16] Thus both Kuhn and the Starnberg group can correctly be described as proposing 'internalist' account of intellectual change in the 'mature' sciences.

[17] Although J. Law, 'The Development of Specialties in Science: the case of X-ray crystallography', *Science Studies*, 3 (1973), 275–303 does suggest there are at least three types of specialties in the modern sciences.

[18] See, for instance, the discussion in D. Edge and M. Mulkay, 'Fallstudien zu wissenschaftlichen Spezialgebieten', in N. Stehr and R. König (eds.), *Wissenschaftssoziologie*, Köln: Westdeutscher, 1975.

[19] In the sense of following the reorientation of much work following the diffusion of Kuhn's essay.

[20] As proclaimed by, *inter alia*, D. Bloor, *Knowledge and Social Imagery*, London: Routledge & Kegan paul, 1976, ch.1 and H. Collins, 'Stages in the Empirical Programme of Relativism', *Social Studies of Science*, 11 (1981), 3–10.

[21] See, for example, S. B. Barnes and S. Shapin (eds.), *Natural Order*, London: Sage, 1979; H. M. Collins, 'The Seven Sexes: a study in the sociology of a phenomenon, or the replication of experiments in physics', *Sociology*, 9 (1975), 205–24; K. Knorr *et al.* (eds.), *The Social Process of Scientific Investigation*, Sociology of the Sciences Yearbook 4, Dordrecht:

Reidel, 1980; B. Wynne, 'C. G. Barkla and the J Phenomenon', *Social Studies of Science*, 6 (1976), 307–47. For a critical discussion of some of this work see R. Whitley, 'From the Sociology of Scientific Communities to the Study of Scientists' Negotiations and Beyond', *Social Science Information*, 22 (1983), 681–720.

[22] As in B. Latour and S. Woolgar, *Laboratory Life*, London: Sage, 1979.

[23] Ibid., ch. 6.

[24] As characterized by K. Knorr-Cetina, 'Scientific Communities or Transepistemic Arenas of Research?' *Social Studies of Science*, 12 (1982), 101–30.

[25] Which is what Latour and Woolgar, op. cit., 1979, note 22 and K. Knorr—Cetina, *The Manufacture of Knowledge*, Oxford: Pergamon, 1981 seem to do. Compare R. Whitley, op. cit., 1983, note 21.

[26] For a useful discussion of the idea that complexity prevents a naturalistic social science, see D. Thomas, *Naturalism and Social Science*, Cambridge University Press, 1979, pp. 13–17.

[27] This is not to say that ontological aspects are irrelevant to differences between the sciences but simply to point out that the particular ways in which research topics are conceived and connected are historically contingent and variable. Thus disciplines and specialisms are units of social organization which structure the production of intellectual innovations in particular ways which can, and do, change in different contexts.

[28] Compare E. Mendelsohn, 'The Emergence of Science as a Profession in Nineteenth Century Europe', in K. Hill (ed.), *The Management of Scientists*, Boston: Beacon Press 1964.

[29] For a similar view, see R. Collins, *Conflict Sociology*, New York: Academic Press, 1975, p. 492.

[30] Especially the constituent fields of natural history, and especially in Britain. See, for instance, D. E. Allen, *The Naturalist in Britain, A Social History*, London: Allen Lane, 1976; P. L. Farber, *The Emergence of Ornithology as a Scientific Discipline, 1760–1852*, Dordrecht: Reidel, 1982; R. Porter, 'Gentlemen and Geology: the Emergence of a Scientific Career, 1660–1920', *The Historical Journal*, 21 (1978), 809–36.

[31] As R. S. Turner has emphasized. See his PhD thesis entitled *The Prussian Universities and the Research Imperative, 1806–1848*, Princeton University, 1972, pp. 387–99, and his 'The Prussian Professoriate and the Research Imperative, 1790–1840', in H. N. Jahnke and M. Otto, *Epistemological and Social Problems of the Sciences in the Early Nineteenth Century*, Dordrecht: Reidel, 1981.

[32] W. O. Hagstrom, *The Scientific Community*, New York: Basic Books, 1965, ch. 1; N. Storer, *The Social System of Science*, New York: Holt, Rinehart & Winston, 1966.

[33] J. R. Ravetz, *Scientific Knowledge and Its Social Problems*, Oxford: Clarendon Press, 1971, ch. 3.

[34] Notably Latour and Woolgar, op. cit., 1979, note 22 and Knorr-Cetina, op. cit., 1982, note 25.

[35] Hagstrom, op.cit., 1965, note 32, ch. 1. Compare R. Whitley, op.cit., 1972, note 2.

[36] These, and similar, points are elaborated in R. Whitley, op. cit., 1983, note 21.

[37] Compare B. Latour, 'Give me a Laboratory and I will Raise the World', in K. D. Knorr-Cetina and M. J. Mulkay (eds.), *Science Observed*, London: Sage, 1983. This sort of historial myopia is not uncommon among the recent studies of scientists' behaviour in laboratories.

[38] Pantin has characterized the field sciences as 'unrestricted' while physics and chemistry are 'restricted'. See C. F. A. Pantin, *The Relations Between the Sciences*, Cambridge University Press, 1968, ch. 1. This distinction has been linked to differences in the organization and control of research in R. Whitley, 'The Sociology of Scientific Work and the History of Scientific Developments', in S. S. Blume (ed.), *Perspectives in the Sociology of Science*, New York: Wiley, 1977. See also A. Rip, 'The Development of Restrictedness in the Sciences', in N. Elias *et al.* (eds.), *Scientific Establishments and Hierarchies*, Sociology of the Sciences Yearbook 6, Dordrecht, Reidel, 1982.

[39] Compare Ravetz, op. cit., 1971, note 33, ch. 4.

[40] See, for example, P. R. Lawrence and J. W. Lorsch, *Organization and Environment*, Harvard University Press, 1967; L. Karpik (ed.), *Organization and Environment*, London: Sage, 1978, chs. 1, 4, and 5.

[41] On the contexts of changing industrial and organizational structures, see A. L. Stinchcombe, 'Social Structure and Organization' in J. G. March (ed.), *Handbook of Organization*, Chicago: Rand McNally, 1965; A. D. Chandler, *Strategy and Structure*, MIT Press, 1962; N. Kay, *The Evolving Firm*, London: Macmillan, 1982.

[42] P. DiMaggio and P. M. Hirsch, 'Production Organizations in the Arts' in R. A. Peterson (ed.), *The Production of Culture*, London: Sage, 1976 p. 79. Compare T. S. Kuhn, *The Essential Tension*, Chicago University Press, 1977, ch. 9.

[43] R. Collins, op. cit., 1975, note 29. pp. 486–91.

[44] See C. E. McClelland, *State, Society and University in Germany, 1700–1914*, Cambridge University Press, 1980, chs. 4 and 5.

[45] One of the most notorious cases is that of Lord Rayleigh who submitted a paper anonymously to the British Association which was rejected only to have it accepted with some embarrassment when he informed the committee who the author was. Cf. R. Merton, 'The Matthew Effect in Science' in *The Sociology of Science*, Chicago University Press, 1973. For a discussion of how metropolitan élites controlled contributions to British Association meetings in the 1830s and thus the public image of science, see J. Morrell and A. Thackray, *Gentlemen of Science*, Oxford University Press, 1981, ch. 6.

[46] A. Stinchcombe, 'Bureaucratic and Craft Administration of Production', *Administrative Science Quarterly*, 4 (1959), 168–87.

[47] Eccles's criticism of Stinchcombe's analysis of the construction industry does not affect the contrast being drawn here. See R. G. Eccles, 'Bureaucratic vs. Craft Administration: the relationship of market structure to the construction firm', *Administrative Science Quarterly*, 26 (1981), 449–69.

[48] There is some dispute about whether a system of salaried posts for researchers did exist in France at the end of the eighteenth century. It seems to be partly a matter of terminology and partly a matter of numbers. See, for example, H. Gilman McMann, *Chemistry Transformed, the paradigmatic shift from Phlogiston to Oxygen*, Norwood, New Jersey: Ablex Publishing Corp., 1978; R. Hahn, 'Scientific Careers in 18th Century France', in M. Crosland (ed.), *The Emergence of Science in Western Europe*, Macmillan, 1975; R. Fox, 'Science, The University and the State', in G. Geison (ed.), *Professions and the State in France*, University of Pennsylvania Press, 1984. On the distaste of English 'gentlemen of science' for salaried employment of researchers in the early nineteenth century, see J. Morrell and A. Thackray, op. cit., 1981, note 45, pp. 322, 423–4, 462.

[49] Although they were very useful supplements to other sources of income. See E. Crawford, 'The Prize System of the Academy of Sciences, 1850–1914', in R. Fox and G. Weisz (eds.) *The Organisation of Science and Technology in France, 1808-1914*, Cambridge University Press, 1980.

[50] There is an extensive literature on the management of research in industry and a number of specialist journals focusing on this topic. Formal planning models are now not as highly regarded as they were ten or fifteen years when optimizing algorithms were thought appropriate.

[51] See, for example, D. J. Hickson and M. W. Thomas, 'Professionalisation in Britain', *Sociology*, 3 (1967), 37–54; H. Wilensky, 'The Professionalisation of Everyone', *American Journal of Sociology*, 59 (1964), 137–57, among others. Terence Johnson makes some critical remarks on this literature in his *Professions and Power*, London Macmillan, 1972.

[52] M. Sarfatti Larson, *The rise of professionalism*, University of California Press, 1977, pp. 14–17 and *passim*.

[53] Randall Collins, op. cit., 1975, note 29, p. 341.

[54] As in the case of Davy, Dalton, and Faraday among others, see M. Berman, *Social Change and Scientific Organisation*, Heinemann, 1978, and R. H. Kargon, *Science in Victorian Manchester*, Manchester University Press, 1977; compare C. A. Russell *et al.*, *Chemists by Profession*, Open University Press, 1977, ch. 6. However, Gustin points out that many of the successful apothecary training schools were linked to research journals and leading chemical researchers in Germany around the 1790s and 1800s; see B. H. Gustin, *The Emergence of the German Chemical Profession, 1790-1867*, University of Chicago unpublished PhD thesis, 1975, pp. 62–76.

[55] That is, labour markets for scientific skills developed which were dominated by scientific reputations.

[56] M. Berman, 'Hegemony' and the Amateur Tradition in British Science', *Journal of Social History*, 8 (1975), 30–50.

[57] Collins, op. cit., 1975, note 29, p. 341.

[58] For a discussion of Avery's concern with checking results before sending papers for publication, see Rene Dubos, *The Professor, The Institute and DNA*, Rockefeller University Press, 1976. The 'Summerlin affair' demonstrates the dangers of claiming important results which cannot be replicated, see J. Hixson, *The Patchwork Moue*, New York: Doubleday, 1976.

[59] Competition is not simply for personal recognition but for domination and control of others' research. Hence, its form, scope, and intensity affect the development of ideas and results much more directly than the recognition-exchange model of scientific communities suggests.

[60] Collins, op. cit., 1975, note 29, p. 478.

[61] See the papers by Harvey, Pickering and Pinch in K. Knorr *et al.* (eds.), *The Social Process of Scientific Investigation*, Sociology of the Sciences Yearbook 4, Dordrecht: Reidel, 1980 and by Collins and Pinch in H. M. Collins (ed.), 'Knowledge and Controversy', *Social Studies of Science*, 11 (1981), Special Issue.

[62] First of all in classical philology as discussed by Turner, op. cit., 1972, note 31, pp. 305–18.

[63] As is demonstrated by the fate of David Bohm's attempt to challenge the dominant orthodoxy in quantum mechanics in the early 1950s. As Pinch suggests, Bohm was only able to make such a challenge convincingly because of his current high status in theoretical physics and so could not be entirely ignored or discussed as a crank. See T. J. Pinch, 'What Does a Proof Do if it Does not Prove? A Study of the Social Conditions and Metaphysical Divisions Leading to David Bohm and John von Neumann Failing to Communicate in Quantum Physics', in E. Mendelsohn *et al.* (eds.), *The Social Production of Scientific Knowledge*, Sociology of the Sciences Yearbook 1, Dordrecht: Reidel, 1977.

[64] Compare W. O. Hagstrom's model of disciplinary segmentation in his op. cit., 1965, note 32, ch. 4. However, modern physics is rather unusual in its degree of homogeneity and hierarchy and other fields manifest greater intellectual pluralism and sustained conflicts over goals and procedures.

[65] Of course, where alternative ideals and audiences exist which also have considerable social prestige, such as the traditional humanities in many European countries, this domination of the cultural system by the laboratory-based natural sciences is limited. However, new fields seeking resources and legitimacy have tended to follow 'scientific' goals and procedures rather than attempting to establish themselves as key parts of the traditional 'high' culture; not least because they were often engaged upon the secularization and demythologizing of 'tradition' and its controllers.

[66] The key role of these sorts of work procedures in establishing new areas of research is discussed in the following: G. Allen, *Life Science in the Twentieth*

Century, Cambridge University Press, 1978, pp. 61–9; 'The Transformation of a Science: T. H. Morgan and the Emergence of a New American Biology' in A. Oleson and J. Voss (eds.), *The Organisation of Knowledge in Modern America, 1860–1920*, Johns Hopkins University Press, 1979; 'The Rise and Spread of the Classical School of Heredity, 1910–1930', in N. Reingold (ed.), *Science in the American Context*, Washington, D. C.: Smithsonian Institution, 1979; A. Kuper, *Anthropologists and Anthropology*, Harmondsworth: Penguin, 1975, ch. 1; J. Morrell, 'The Chemist Breeders', *Ambix*, 19 (1972), 1–46; R. S. Turner, op. cit., 1972, note 31, pp. 281–319.

2

REPUTATIONAL CONTROL OVER SCIENTIFIC WORK AND THE GROWTH OF EMPLOYMENT OPPORTUNITIES FOR SCIENTISTS

REPUTATIONAL work organizations do not need to control jobs and labour markets to ensure their control over work and practitioners as long as the reputations they control are socially prestigious and lead to material rewards indirectly. In this respect they are different from craft and professional systems of work administration – at least analytically. The sciences existed as such systems of work direction and control before labour markets in scientific skills became established and, of course, reputational means of work control remain important even when practically all scientific work is produced by employees. However, the development of jobs tied to reputations, and of employment organizations oriented primarily to intellectual goals as defined and controlled by practitioners, resulted in substantial changes in the organization and control of scientific work. In particular, the combination of training programmes, certification agency, jobs and facilities in universities was a powerful influence on the structure of the sciences. In this chapter I shall discuss some aspects of this development in the general context of changing relations between employment organizations and reputational work organizations.

SCIENCES AS NON-LABOUR MARKET WORK ORGANIZATIONS

First, we need to outline the general characteristics of the modern sciences as non-labour market reputational work organizations before they controlled jobs. As discussed in the previous chapter, these require some autonomy from dominant groups, a formal symbol system for communicating task outcomes, standardized work procedures and control over their production and certification, and control over socially prestigious and desired reputations. Given these conditions, sciences can function as distinct organizations which control, direct, and evaluate work. However, the extent to which any particular field manifests these characteristics obviously varies and so too does their control over what and how research is done. Where a given reputational group has to share the evaluation of task outcomes with other scientific or non-scientific groups, where techniques are not highly standardized and exclusively controlled, and reputations are not highly valued – both within science and elsewhere – its control over practitioners will be limited.

Before intellectual priorities, procedures, and criteria became strongly entrenched in employment structures and labour markets their stability and rigidity was much more limited than after this development had occurred. In general, the number of practitioners in any one field was fairly low and their intellectual identity was not restricted to a single set of practices and commitments which determined what work they did and how they did it. Who could be termed a 'physicist' in early nineteenth-century, England, for example, was ambiguous and subject to sudden changes.[1] Both mobility between fields and of goals and procedures was commonplace so that intellectual and social boundaries were more fluid and amenable to individual influence than became the case of 'professionalized' science. Knowledge tended to be less objectified in social structures and more connected to individuals than to organizations.[2] French mathematicians in the early nineteenth century, for instance, adopted idiosyncratic styles which were reflected in their textbooks and other publications.[3]

Science was not, in general, so institutionally separated from other forms of intellectual production and validation, nor were the special sciences so distinguished from sciences as a whole or from one another, as they later became. Audiences for research results were relatively undifferentiated and more likely to be personally known to contributors. Reputations could be, and were, awarded across the whole of 'natural philosophy' for major contributions by the general scientific élite. Equally, reputations in some fields, such as geology, were associated with non-scientific criteria and groups such as theologians and philosophers.[4] Skills were not always standardized within a particular field and did not need lengthy training programmes to be acquired. The case of Sedgwick successfully educating himself in Geology after being elected to the Woodwardian Chair is a well-known instance of this point.[5] The continued influence of 'amateurs' in geology and biology throughout the nineteenth century, especially in England, demonstrates the lack of strong control over training and certification exerted by the reputational system in these fields.[6] A further illustration is provided by German mathematics. According to Mehrtens, 'all major mathematicians educated up to 1830 in Germany were essentially autodidacts'.[7] Science was much more open to practitioners from a variety of intellectual backgrounds, then, and to innovations from a variety of sources than it later became.

Because boundaries, goals, and audiences were relatively fluid and varied in the 'pre-professionalized' sciences, contributions tended to be more wide-ranging and general than the typical product of Kuhnian 'normal' science. Reputations were not allocated by a single group of full-time practitioners who were committed to a single set of specific beliefs and practices but by a shifting and diffuse audience of scientists who had much broader intellectual identities and commitments. Consequently, uncertainty about the significance of results was considerable and liable to create major conflicts and disputes such as debates over the foundations of mathematics and the importance of 'pure' mathematics in the early nineteenth century.[8]

Where expertise and competence were not sharply defined and controlled by certification agencies, the boundaries of evaluative groups and criteria could be drawn only weakly and hence were liable to rapid modification and change. Contributions to knowledge thus had to be made in a social situation of considerable uncertainty and variety. In these circumstances, scientists could not take 'paradigms', or other configurations of assumptions and commitments, automatically for granted – or, at least, not to the extent of assuming that reputations would be awarded for minor variations on well-established themes. Instead, many scientists had to produce results which would convince the intellectual élite of their general importance to general intellectual goals rather

than simply to highly narrow and specific ones. The general
nature of most scientific journals before the nineteenth
century is an instance of this point.[9]

Competition between scientists for reputations in these
circumstances tended to be more diffuse and more concerned
with general intellectual goals and priorities than it later
became. To gain high reputations in 'pre-professional' science, practitioners had to show not only that their results were
significant, reliable, and important for particular problems
but also that those problems were important ones for major
areas of scientific concern. This meant that scientists competed for control over a number of cognitive dimensions
including theoretical models and priorities. Because audiences
were relatively fluid and not tied to particular labour markets,
scientists could not compete for specialist reputations from a
definite group with particular commitments and beliefs but
had to obtain reputations from a much more general and
diffuse group of colleagues. They were thus competing with a
relatively variegated set of practitioners pursuing diverse
interests who did not – or could not be assumed to – have a
common training and perspectives. Reputations for major
contributions could, therefore, change as small variations in
membership of the field occurred. As Cannon suggests, the
small numbers involved in research in the early and mid-
nineteenth century meant that the 'prevailing opinion' on a
subject often reduced to four men being for it, two against it,
'two had not strong convictions, and no one else knew enough
about it to have any serious opinions at all'.[10]

The degree of originality in this situation was not as
constrained and tempered by the need to co-ordinate one's
research with that of a definite group of colleagues, who
collectively controlled separate labour markets, as it later
became. While some specialist groups did exist and constituted distinct fields of endeavour, they did not control so
much of the individual's 'life chances' as they later came to.
Their membership was not as fixed by training programmes
and jobs and so their dominant preconceptions and procedures were not so entrenched and rigid as some academic
'disciplines' became. The individual scientist's dependence on
a particular group of colleagues was correspondingly less and

her or his freedom of manoeuvre greater. Originality could, therefore, be more wide-ranging without necessarily leading to rejection and so theoretical pluralism and diversity was relatively high in the 'pre-professional' sciences.

This autonomy from particular reputational groups did not' however, necessarily imply autonomy from the scientific élite as a whole, or from general cultural pressures. The relative fluidity of intellectual boundaries and work skills reduced the salience of particular reputational groups but made individual practitioners more open to influence from the general scientific élite because limited special expertise was required to pass judgement on work done in specific fields. To obtain a reputation for substantial contributions, the worth of these had to be demonstrated to a broadly based audience which could insist on its own general criteria of validation and significance. The role of dominant groups in science as a whole was thus important in legitimating contributions in particular areas, not least through their control of the general and most prestigious journals such as the *Comptes Rendus* and the *Philosophical Transactions*. To gain an audience scientists had to conform with the general validation criteria of the currently dominant scientific élite, and so while they might have some autonomy in the details of their work, they were directly subject to the procedures and preferences of the general reputational group in control of the ideals and practices of sciences as a whole.[11]

Additionally, the degree of standardization and formalization of skills was less marked in the sciences when they did not directly control and segment labour markets. Craft methods of conducting research did, of course, exist and did serve to differentiate areas of scientific work but they did not control access to jobs nor were they institutionalized in separate training programmes. Acquisition of research skills was an *ad hoc* affair and very dependent on personal connections.[12] Consequently results from different countries were by no means simply compared and disputes over technical adequacy were common.[13] Similarly, the communication system was often not so highly structured and formalized as to permit easy and unambiguous transmission of ideas and task outcomes so that personal contact and knowledge were important elements of the collective control of scientific work. Standardizations of error limits in experiments in physics, for instance, did not become widely accepted until the 1820s.[14]

In general, then, 'pre-professional' science was characterized by considerable fluidity of social and intellectual identities, mobility between topic area, generality and diffuseness of skills, diversity of work goals, and breadth of conflict and competition. The degree of control that particular, separate, reputational groups had over the work goals and skills of practitioners was limited and so highly differentiated scientific fields with distinct social and intellectual boundaries were not greatly evident. Instead, the boundaries and goals of the special sciences tended to be weakly institutionalized and subject to influences by particular individuals and patterns of group mobility. Chemistry, for example, did not really develop a distinct identity, subject matter, and method of approach in many European countries until the end of the eighteenth century,[15] and distinctions between what became physics, astronomy, and mathematics changed considerably over the course of the eighteenth century.[16] Individual scientists could, and did, contribute to a variety of fields without undergoing special training and without becoming identified with any particular 'discipline'. Science, or natural philosophy, was the broad area within which reputations were sought and this itself overlapped with philosophy and other areas of cultural production well into the nineteenth century.[17] Specialist groups did, of course, exist but these were not entrenched in labour markets and employment organizations, nor did they control training programmes or monopolize access to communication media - which in any case were relatively undifferentiated and unspecialized. Individual specialists controlled validation processes and access to reputations in general but they did not, I suggest, constitute coherent groups which functioned as distinct organizations that continued to operate and control research in the same way when individual members changed. Research was therefore more personally controlled and variable than it later became as audiences increased anonymity and geographical distance.

As separate reputational work organizations, then, the

special sciences had only a limited and temporary existence before they began to control labour markets and employment organizations. Science in general had sufficient autonomy and control over prestigious reputations to gain the allegiance of practitioners and control task outcomes but its boundaries and nature were only weakly institutionalized and subject to considerable conflict and modification. As long as scientists disputed the basis of awarding general scientific reputations by advocating conflicting claims to the significance of their work which included general metaphysical dimensions and conceptions of science,[18] the likelihood of particular sciences forming distinct social and intellectual groupings which were relatively stable remained low because conflicts over reputations involved meta-scientific criteria and norms, i.e. appeal to standards which were beyond the boundaries of particular fields.

REPUTATIONAL ORGANIZATIONS AND EMPLOYMENT ORGANIZATIONS

The growth of employment opportunities which were clearly dependent on international scientific reputations did not occur very quickly, nor did it alone radically alter the structure and organization of the sciences. However, this move towards 'professionalizing' the sciences did result eventually in nearly all scientific work being done by employees and reputational control over labour markets did bring about substantial changes in the general characterization of science just outlined.[19] Before continuing to discuss the particular processes of academic institutionalization of scientific work and its consequences for the organization of the sciences, there are some general points about relations between reputational organizations and employment organizations which need to be made.

In craft and professional systems of work administration, labour market status is more important than employment organization status. As outlined in the previous chapter, this is also the case in reputational work organizations but here collegiate reputations determine, to a very large degree, labour market status. While training and certification procedures control entry to labour markets, they are less important for obtaining good jobs, promotion, and access to research facilities than collegiate reputations. Because of the constant revision of work skills and communal evaluation of task outcomes in the sciences, initial training and certified competences have less impact on what work is done and how it is done than they do in other fields of professional work. While some employers may only require the basic skills for their goals, and so may ignore current reputations, those concerned to produce public knowledge will rely on these reputations to ensure continued production of valid and significant contributions. For these latter employers, certification of basic, 'disciplinary' skills will be only a minimum requirement for employees while current capabilities as assessed by reputations gained from previous work will be more important in making employment and promotion decisions. Labour market status as defined by skill certification agencies is less important here than labour market status as defined by current collegiate reputations. Because such reputations change over individuals' lifetimes, labour market status similarly alters in a way which seems much less important in other professional work organizations where, for most practitioners, skills acquired during the initial training phase suffice to define labour market status in a semi-permanent manner. This aspect of reputations in the sciences makes practitioners insecure and heightens competition for high reputations when jobs are scarce.[20]

During the nineteenth century, the growing dependence of livelihoods on labour market status as defined by collegiate reputations both made such reputations more crucial to scientists than before and institutionalized intellectual commitments and practices in employment organizations. The pursuit of collegiate reputations became more intense as more depended on them, and the entrenchment of the beliefs and boundaries of reputational groups in employment organizations reified and reinforced intellectual and social hierarchies. Individual dependence upon colleagues increased when employment status was derived from collegiate reputations and this increased mutual dependence between specialist colleagues made it less likely that existing beliefs and practices were seriously challenged by practitioners because dominant commitments had become occupational identities. This pressure towards intellectual conformity was especially strong where reputational communities totally controlled access to jobs and facilities so that individuals were highly dependent on a single, coherent audience for reputations which in turn determined employment status.

In practice, however, employers have rarely been so totally dominated by a single, cohesive reputational group. Instead, the extent to which their goals and policies have been controlled by scientists seeking collegiate reputations has varied considerably between the sciences, e.g. between philosophy and mathematics, and in different historical periods, e.g. between nineteenth and late twentieth-century philosophy. In considering how the development of employment for researchers has affected the organization of scientific fields, these variations are obviously important factors to be taken into account and require some discussion before dealing with the emergence of university-based disciplines as major units of labour market organization in the sciences.

In general terms, relations between employers and reputational groups can be discussed on three major dimensions. The first of these is the degree to which employers are oriented to the production of public knowledge or just buy particular knowledge production skills for private purposes. The second dimension refers to the influence of collegiate reputations in general on personnel policies and organizational status while the third covers the extent to which such influence is exercised by one or two reputational groups which are internally cohesive and dominated by a particular élite.

These three dimensions all bear on the extent of control that reputational organizations exert over the sort of scientific work that is carried out in employment organizations, and how it is carried out. Depending on the position employers take on these dimensions, sciences vary in structure from the ideal case of Kuhnian paradigm-bound communities, where dominant reputational groups monopolize the evaluation of results and such evaluations dominate the allocation of jobs, access to facilities and to senior posts in employment organizations, to the craft or bureaucratic-professional cases where sciences are responsible for training and certifying particular skills for a largely anonymous labour market, but have only very limited control over how those skills are used, or for what purposes, as in the cases of accountants and engineers.

The first dimension affects the possible positions of employment organizations on the other two since, if employers are not oriented to the public system of knowledge validation and assessment, it is unlikely that administrative hierarchies will be governed by the public reputational system[21] nor that particular reputational groups will affect research strategies. Equally, of course, the second dimension dominates the third. Thus by combining these three dimensions we can distinguish between three major situations in which research is conducted by employees.

First, there is the situation where employers' goals are quite separate from those of public science groups and scientists are hired for private purposes. Little or no feedback to reputational systems occurs in these cases except through the expansion of job opportunities for scientists with particular skills affecting the size and importance of particular training programmes. In so far as trainee scientists provide the bulk of scientific labour power in the public sciences, this factor affects knowledge production in these fields.[22] Reputational systems have little or no impact on what work is done, how skills are combined and managed, or how results are evaluated in this situation. Individual dependence on reputational groups is low and employment status usually more important than reputational and labour market standing. Given that employer priorities diverge considerably from those of reputational organizations, research is unlikely to fit in with the particular boundaries and programmes of the latter. Rather, scientists will tend to study topics which are irrelevant to reputational considerations and/or which require skills, knowledge, and procedures from a number of intellectual fields and so transcend reputational boundaries. This situation is typical of much industrial science.

The second situation includes a wide range of cases where employers' goals partly overlap with those of reputational communities and partly derive from non-scientific objectives such as medical, military, or social ones. The expansion of such 'goal-directed' research[23] since the 1950s has had radical effects in many parts of the sciences with traditional boundaries and goals being broken down and transformed into highly fluid and changeable social and intellectual structures. These research organizations are often strongly oriented to reputational systems in that they employ scientists with particular skills as certified by training agencies and rely on reputational criteria in making promotion decisions. However, they also have to fulfil other objectives and direct their work along non-traditional lines. These multiple, sometimes conflicting goals, result in skills being combined in novel ways for new intellectual goals so that results cannot easily be fitted into existing boundaries and reputational priorities. Given scientists' primary orientation to reputations as the means of controlling and validating work this has resulted in new reputational groups developing around what might be called 'mixed' goals. The growth of employment opportunities in such 'hybrid' organizations has inevitably meant a weakening of traditional boundaries and of the control exerted over scientific research by traditional 'disciplinary' groups. Depending on how strong are traditional hierarchies of scientific prestige, and the widespread availability of research resources, control over research priorities and the assessment of the significance of results by established élites has been considerably reduced by these developments in some sciences. This situation is typical of much state-supported research conducted in full-time research labouratories funded by central agencies. It can be summarized as producing 'state science'.

The third situation is the traditional ideal of the scientific community where employers' goals are identical, and subservient, to those of reputational groups. Here, recruitment and promotion decisions are made on grounds of reputations gained from national and international groups of scientists which validate and assess the worth of results and a major goal of employment organizations is to produce the best knowledge in the field(s). Objectives are not necessarily single in the sense of being oriented solely to one reputational field but are wholly within the public scientific set of knowledge goals. Control over what work is done, how it is conducted, and how its significance is evaluated within employment organizations, follows the imperatives and criteria of reputational organizations in this situation. Employment organization status is derivative from reputational and labour market standing. This situation can be characterized as typical of 'academic' science.

TABLE 2.1

Reputational Control of Employers' Research Goals and Types of Science

Reputational control of employers' goals and policies	Types of Science		
	Academic science	State science*	Industrial science
Degree to which employers are oriented to public science goals	High	High to Medium	Low
Degree to which employers follow reputational criteria and values in their personnel policies	High	Medium	Low
Degree to which promotions and other organizational rewards are governed by a particular reputational élite	High to Medium	Medium to Low	Low

* Excluding military research and other secret work but including combined state and voluntary organization supported fields.

These three situations can be summarized as in Table 2.1. Here we can see how industrial science is relatively independent or reputational groups and values – except perhaps when hiring graduates of the 'best' educational establishments – and simply buys particular knowledge production skills for private purposes. State science, on the other hand, is more oriented to public science goals and tends to follow reputational critera in its recruitment and promotion policies. However, the multiple, and sometimes contradictory, goals pursued by such employers mean that non-scientific criteria will also affect personnel policies. Many bio-medical research laboratories, for example, have qualified medical doctors as their directors, even when they have never practised clinically and the research has no direct connection with clinical medicine. In fact, in sciences with numerous employment opportunities and multiple goals, reputational priorities may be influenced more by employers' policies than vice versa as directors make recruitment and promotion decisions to suit their objectives which direct research in certain ways so that reputational communities form around these topics and techniques. Control over resources here can lead to direction of research, especially if similar organizations follow the same policy. High administrative authority thus influences reputational prestige as particular areas and techniques are seen as crucial to solving the 'cancer problem', for instance, and so workers in these fields are accorded greater status and recognition.

From this it follows that state research organizations are unlikely to produce knowledge which contributes to only a single field unless that field is simply the distinct area of research formed by the results of scientific work conducted in all laboratories of that type. For example, it could be argued that the new field of oncology arose only because new employment organizations were established to produced knowledge which would enable cancerous diseases to be controlled and this area is composed of research outcomes from these organizations only. Even here, though, research from such laboratories contributes to a number of fields, many of which overlap, and the study of cancerous growths is also undertaken in many research organizations. In other words, in the bio-medical sciences, and many others, employment organizations contribute to a variety of subject areas and each field acquires contributions from a variety of employment organizations. Relations between employment and reputational organizations are here multiple and variable with common boundaries being rather rare.

In academic science, where employers are more directly oriented to reputational goals, appointment to senior posts are more likely to follow reputations, but the possible plurality of relations between employment organizations and reputational communities means that administrative hierarchies will not always match prestige hierarchies in scientific fields. Indeed, given the semi-permanent nature of employment statuses in most scientific work organizations these may determine the criteria by which work is assessed as being especially significant, and so academic establishments may dominate reputational communities as Elias has recently argued.[24] While such domination may arise in some fields, internal differentiation and the plurality of links already mentioned make it unlikely to occur in general. Instead, the extent to which controllers of reputations in a particular field are also controllers of jobs, promotions, and access to research facilities in that field needs to be empirically established and the conditions under which variations in such control occur investigated. Obviously a wide availability of funds, jobs, and communication media will reduce tendencies to monopolistic domination just as theoretically unified and resource concentrated fields will tend to encourage it.

Scientists in most employment organizations thus develop a certain degree of autonomy from the dominance of any single reputational group. Even in highly centralized and unified fields, such as physics and economics, practitioners attempt to achieve some independence from dominant groups by developing specialized skills and focusing on novel topics and concerns. Often this is achieved through colonizing other fields and showing how they can be better understood by using different techniques and conceptual approaches to those current in that area. These moves to greater autonomy on the part of some scientists in such 'disciplines' may be countered by attacks on 'applied' sub-fields as not being 'real' physics or economics from the established élite and by refusal to publish much of the work done in these areas in the top journals. In any event, the domination of all the research carried out in certain employment units by a single reputational group is only one type of relation between employment and reputational organizations and others are empirically feasible in sciences where employers are largely oriented to reputational goals. Here, too, there is no reason to assume that goals and boundaries of the two types of organizational ways coincide and are identical. Rather a plurality of possible links and overlaps occur in different sciences in different circumstances with different consequences.

It can be seen from this discussion that relations between employers and reputational organizational can be summarized in terms of the relative influence that the latter can exert on the research policies and reward systems of the former. In areas like elementary particle physics this influence is quite high and is exerciese by a relatively cohesive international élite, while in many bio-medical research areas laboratory directors have much greater discretion from any single scientific élite in determining their research strategies and recruitment policies. However, not all academic departments are necessarily dominated by a single reputational élite, nor are all state-supported fields as independent of particular reputational groups as many of the bio-medical sciences are. University departments of sociology in many countries, for instance, are probably less controlled by a single élite group since 1960 than is contemporary plasma physics research conducted in state laboratories for fusion reactor physics. Furthermore, the sheer size of state investment in research since the Second World War has altered academic structures and relationships between employment units such as university departments and reputational fields. Old-established fields which were based upon departments and institutes in universities have become internally highly differentiated and have altered their goals and methods so much that departmental titles bear little relation to the research straregies being pursued by employees. This has been especially noticeable in the biological sciences since the invasion of work methods and apparatus from chemistry and physics.[25]

The modern public sciences, then, exhibit varied relationships between employers and reputational organizations and include fields dominated by researchers employed by state, and other central, agencies. Intellectual production is not restricted to university-based disciplines, nor do skill training and certification units necessarily dominate the setting and evaluation of research strategies. However, it was the combination of training, employment, and the pursuit of national and international reputations in universities which produced the dominant unit of knowledge production in the nineteenth and early twentieth centuries – the academic discipline. This combination structured the context in which present-day scientific fields emerged and became established and still dominates research in many areas. By systematically connecting organizational status and authority to extra-local reputations for contributions to collective intellectual goals, the nineteenth-century university system bureaucratized intellectual production and organized it into distinct, specialized disciplines. These disciplines were both units of labour market definition and control, and of intellectual production and validation.

THE ENTRENCHMENT OF REPUTATIONAL WORK ORGANIZATIONS IN UNIVERSITIES

Apart from the small number of royal academics established in the seventeenth and eighteenth centuries to provide pensions and other forms of financial and instrumental support for natural philosophers, the first major employer of scientists for knowledge production was the reformed Prussian university system.[26] Certainly this system provided the model for many subsequent reforms of higher education and provision of resources for scientific work, to the extent that it has become seen as 'natural' and 'normal' for science to be located in universities and the common model of professionalized science assumes the dominance of university employees.[27] As the major source of employment for scientists primarily oriented to public reputational goals, university departments and institutes are obviously key organizations for any analysis of relations between reputational and employment organizations.

As employment organizations for scientists pursuing collegiate reputations through publication or research results universities are differentiated from other research-oriented employers, such as the Max-Planck Institutes and the CNRS, by their commitment to teaching goals and their certification and assessment role. Even the most research-oriented university has to educate some students, usually at both the undergraduate and post-graduate levels, and to arrive at public judgements of their merits. While the development of the research degree enabled scientists to conduct research at the same time as training students, universities have always had to educate and assess students for non-professional purposes, that is for non-scientific ends, and so cannot be simply reduced to convenient institutions for locating 'pure' scientists. They are multi-purpose organizations and pursue a number of different goals which are not always compatible with the pure science ideal. As such they resemble the 'hybrid' employment organizations mentioned above rather more than is sometimes allowed. Furthermore, given that universities existed before the modern sciences became fully institutionalized as separately organized activities, and that they often paid little attention to the production of new, original knowledge as a criterion of appointment and advancement,[28] the inclusion of science in established universities, and the importance of international reputations among scientific colleagues for promotion, did not develop without repercussions for reputational systems. The entrenchment of the modern sciences in university employment units, and the orientation of university policies to international reputations for original contributions to knowledge goals as defined by a group of scientists who dominated the field, undoubtedly altered the ways universities operated. They also, though, affected the organization and ways of working of the sciences.

One of the major consequences of institutionalizing the modern sciences in the reformed university system was the incorporation of science into national pedagogic systems. While this enabled scientists to extend their control of recruits and training programmes into the secondary school system, thus to influence general public images of science, it also ensured that the training and recruitment of scientists became organized in a systematic way for the first time and that this followed already established patterns. Science became established as an activity which deserved not just acceptance as a legitimate pursuit, but active support and facilities from dominant institutions concerned with knowledge transmission and validation in a particular context which ensured that science would fit in with dominant values and goals. The general social functions of the universities were extended to the sciences and scientific ideals became similar to those governing the higher educational system, such as 'neo-humanism' in Germany.[29] This tended to lead to an increased emphasis on 'high' or 'pure' science at the expense of applied or technological concerns as the nineteenth-century universities focused on the provision of education in high culture for a small, privileged élite rather than the training of useful technical experts.[30] In other words, the resources required to develop new directions became very large once communities were institutionalized in universities because that meant creating new training programmes, curricula, and posts. New intellectual fields had to become new disciplines and departments if they were to develop as sustained research traditions and programmes; the dominance of academic departments as the primary way of organizing science resulted in new sciences being forced to become departments if they were to reproduce themselves as reputational communities. This has, of course, been especially observable in the human sciences.

In combining the traditional functions of universities – the inculcation of received wisdom, knowledge, and correct beliefs and modes of behaviour in the privileged young – with the pursuit of new knowledge and training new practitioners in research, the reformers of the Prussian and other universities assimilated the sciences to the dominant ideals and élite groups at the same time as transforming the universities and reasserting state control over them.[31] The universities continued to develop the 'whole man' in certain élite individuals and provide members of the traditional professions and higher civil service, but they now did this through advanced research and the search for truth rather than through Latin disputations.[32] Rather than providing avenues for upward social mobility, as perhaps was the case for a time in France,[33] or useful inventions, science in the German universities became part of a general ideology which served to reinforce the position of existing élites and the 'mandarin' stratum in the civil service. While this usage of the universities was scarcely novel in Europe, its connection to the production of intellectual innovations was.

The reorientation of the German universities in the nineteenth century to the pursit of original scholarship provided a strong organizational base for systematic and sustained research. By legitimating, and providing resources on a considerable scale for, large-scale scientific research, the reforms lead to the growth of the professional scientific community and the domination of that community by German research schools. Even though the gentlemanly 'amateur' tradition continued to produce major figures and contributions in England,[34] the sheer number of practitioners in the German universities coupled with their use of students on a large scale for research, which was in part subsidized by state support for labouratories,[35] led to European science becoming dominated by academic professionalised science.[36]

A second major result of this growth of professionalism, in the sense of jobs depending upon reputations, was that science became 'normalized'. Since research was increasingly carried out by students, reliable means of obtaining results which could be compared and integrated became very important. Inexperienced neophytes could not be expected to contribute to knowledge where task uncertainty was very high because techniques produced variable results which required great experimental and interpretative skill for their meanings to be elucidated. Research schools relying extensively on student labour, then, required fairly simple and reliable technical procedures if they were to make significant contributions to knowledge. As Morrell points out, Liebig's apparatus for the combustion analysis of organic compounds was an essential prerequisite for his successful domination of organic chemistry;[37] it reduced task uncertainty sufficiently for his students to analyse compounds on a large scale and transformed the field. The example of the Saxon scientists in the early nineteenth century is similarly instructive. Jungnickel[38] emphasizes the extensive collaboration between senior scientists required to develop precise and reliable methods of experimental investigation in the natural sciences. This collaboration was largely accomplished through the Saxon Society of Sciences rather than in the Universities of Leipzig and Jena. After precise and reliable methods and instruments had been developed the pattern of research collaboration changed; as Jungnickel says 'methods and instruments that made possible the participation of relative inexperienced helpers such as advanced science students . . . tended to make superfluous the co-operation between highly skilled professionals that had been essential to research earlier'. The professionalization of science in universities, then, tended to emphasize technical precision and reliability and the formulation of problems which could be dealt with by these techniques. Task uncertainty was reduced to a level where neophytes could produce useful results.

By organizing research around the training process, and developing formulations and procedures which permitted trainees to produce valid knowledge claims, the academic professionalization of science developed both a hierarchical structure for conducting scientific research and a style of research which enabled original knowledge to be produced by such hierarchies. In contrast, the dominant mode of knowledge production in pre-professional science – or 'amateur' science – seems to have been more akin to literary and philosophical production than to modern professionalized normal science. Individual craftsmen, or 'geniuses', worked in splendid isolation on relatively diffuse and general problems which varied in conception and preferred approach across practitioners. Occasionally supported by assistants or disciples, these gentlemen amateurs[39] frequently abhorred notions of science as a profession and preferred to work on their own rather than organize research groups with a systematic programme of research and 'rational' division of labour. While they may have competed for reputations among themselves in terms of solving the major problems of the field – and this involved being able to specify those problems as being major – they did so largely as equals and as individuals rather than as schools of knowledge producers entrenched in employment structures. Because training was not systematically organized around employment posts, which monopolized access to research facilities and communication media, it occurred in a largely haphazard and *ad hoc* manner and thus did not impinge greatly upon the form or direction of research. Neophytes may have acquired the basic skills necessary to make valid contributions by working with an 'acknowledged master' but they did so as individuals, usually limited in number, and not as part of the organization of knowledge production. Collaboration between scientists seemed to be relatively uncommon and when it did occur was on the basis of equality rather than as members of a hierarchically structured team, as the above quotation from Jungnickel indicates.

The crucial innovation of Liebig and other producers of knowledge producers was, of course, to change this system radically and to incorporate the training of recruits systematically into the practice of doing research.[40] Henceforth the attraction of a steady flow of research students became a major factor in the development of dominant research programmes in some sciences, and the nature of the problems considered and procedures appropriate for their solution changed accordingly. Given the norm of originality, a relatively formal system of communication and the importance of reputations for jobs, the professionalization of the sciences would probably have resulted in some of these changes anyway but the linking of the training of recruits to the production of new, valid knowledge in the universities accelerated and reinforced the move to organized division of tasks and co-ordination of outcomes by work-place supervisors. Specialization of topic areas and skills among practitioners became echoed and reinforced by specialization and differentiation within research laboratories which were integrated and controlled by an administrative hierarchy organized on the joint basis of technical expertise and knowledge and academic rank. Problems which required collaboration between highly skilled equals or which were relatively diffuse and general so that they could not easily be divided up into separate parts and tasks were less likely to be taken up and worked on in these systems of work organization and control than in more 'amateurish' systems.[41]

The systematic organization of scientific research in employing institutions with clear hierarchies of authority based on differences in expertise and knowledge demonstrated how the arcane and esoteric activity of discovering the laws of the universe, undertaken by a few exceptionally gifted individual 'geniuses', could be rationally administered and controlled. It also showed how discoverers could be produced on a relatively large scale so that when science became seen as useful knowledge, and as a method for producing useful knowledge rather than as a system of fixed knowledge,[42] investors did not need to wait for 'geniuses' to appear but could simply expand the existing system of scientist production. In a sense, the academic domination of science through the production of knowledge producers, and their systematic use as research workers solving differentiated tasks which were co-ordinated by the laboratory head, provided the model of research organization for employers in other spheres. By showing how scientists could be trained and how neophytes could be organized to produce knowledge, the development of research schools in universities demonstrated not only that science could be useful as systematic, fixed knowledge but also that the actual process of knowledge production could itself be planned and organized. Rather than waiting for the fruits of a few individual geniuses to fall off the tree of knowledge and then be applied to extra-scientific problems, the organization of research schools demonstrated how scientific work could be administered as a process oriented to a variety of goals. It was thus capable of being rationally organized and, by extension, planned for goals other than purely academic ones. Sciences as a system of knowledge about the world which was stable, true, and coherent became transformed into a process or method of knowledge production which could be organized and planned. The academic professionalization of science thus paved the way for the large-scale use of science in non-academic environments and for non-academic purposes.

A third major consequence of the institutionalization of the modern sciences in universities was the emergence of powerful academic establishments. By combining the control of training of recruits, curricula in the schools, the organization of research workers in pursuit of research programmes, jobs dependent on reputations based on contributions to knowledge goals and the increasingly expensive and complex facilities needed to make those contributions, university departments and their controlling establishments dominated the sciences. While university employees did not monopolize the production and validation of scientific knowledge in all countries in the nineteenth century, academics did come to hold dominant positions in scientific 'communities', even in countries where the 'amateur' tradition was still a powerful ideological force,[43] and the academic model of professionalization became the 'obvious' means by which reputational communities could obtain control over jobs and research facilities and exclude outsiders. The increasing domination by academics of the sciences, both established and emerging, meant that scientific establishments became academic ones. It also resulted in distinctions between the sciences becoming matters of academic concern and institutionalized in academic boundaries.

Academic establishments were, and are, powerful in the sciences not only because of their control over the production and training of the labour force, a feature they share with other professional establishments, but also because of their domination of the knowledge production process through the extensive use of advanced students as research workers. By monopolizing the inculcation and certification of skills required to carry out scientific research, so that it became impossible for people who had not been through an academic training programme to make 'competent' contributions, academics control entry to the labour market and the meaning of the skills which are 'scientific'. Furthermore, by using students to undertake research tasks which are co-ordinated into a research programme, academics become the major knowledge producers in the sciences because they control more resources. While not all research programmes are equally 'successful', and not all sciences are equally programmable, academic establishments are more able to define what is 'successful' because they control the writings of the 'official history' of the sciences in textbooks and their use in training new recruits. Also, because they control the bulk of the scientific labour force, both staff and students, they are more able to dominate interpretations of the major goals of a science than are individual 'amateurs'. As well as controlling the labour market they dominate the definition of tasks and co-ordination of task outcomes through their control of the major part of the knowledge production system.

Control over a substantial part of the scientific labour force is not the only means of control over knowledge production exerted by academic establishments. As part of the process of increasing precision, purity, and standardization of tasks and technical procedures, which was encouraged by the use of students in research, the technical apparatus required for undertaking scientific research to produce competent contributions became more complex and expensive.[44] The raw materials similarly became more refined and restricted in availability so that the individual practitioner found it more difficult to conduct scientific research without collective resources.[45] The technical means of producing scientific knowledge increasingly outgrew the capacity of individual provision and control so that they had to be collectively organized and controlled. The more concentrated they were in universities and under academic control, the more academic establishments dominated reputational communities. The reduction of task uncertainty through increased precision and standardization of measurements and observations rendered science more esoteric and professional, and more under the control of academics.

The standardization of materials and procedures in many sciences in the nineteenth century was part of the general process of professionalization of scientific work. It aided the academic domination of the sciences through enabling trainees to make contributions to knowledge while distancing the objects and procedures of science from everyday, lay concepts and substances. The simultaneous transformation of scientific research into an activity involving systematic use of esoteric standard techniques conducted by teams and groups in a relatively extensive division of labour and into an activity necessitating collective, and collectively organized, resources which were controlled by a single major institution, the university, was crucial part of the development of professionalized academic science. The standardization and formalization of much scientific research was also important in the development of professional reputational communities. As Kuhn points out,[46] the mathematization of a science enables, *ceteris paribus*, controversies to be speedily resolved and 'normal' science to progress smoothly. By standardizing technical procedures and symbol structures, communication across social and spatial boundaries is facilitated and the reputational community more able to control work practices and outcomes. The reduction of what might be termed technical task uncertainty enabled practitioners with a common set of relatively standardized techniques to communicate task outcomes quickly and easily and so co-ordinate them. Competence in such fields then becomes defined as competence in the use of these techniques for dealing with appropriate problems. Relatively clear social and cognitive boundaries can be demarcated and reified and strong collective social control over the details of work processes exerted. While sciences existed as reputational organizations beforehand, it was their identification with training and employment units built around research programmes using standard procedures which strengthened their identities and power. Equally, the domination of academics over employment and training units which inculcated and applied these procedures facilitated their control over reputational organizations to the extent that scientific fields have become synonymous with university-based 'disciplines'.

This degree of university dominance of the sciences rests upon two main conditions. The first is that university employment remains the major, if not the only, opportunity to gain a livelihood through contributing to knowledge goals or, where alternative sources of employment exist, university status is much higher. The second condition is that the major units of university employment correspond to and dominate those of knowledge production and validation. Where the state or other agencies provide resources and posts for conducting research oriented primarily to public reputational goals, the power of university élities to direct research fields and impose their evaluation criteria on practitioners is correspondingly reduced. Indeed, in fields where full-time research laboratories control the expensive facilities needed for research, or where they provide a substantial number of employment posts, scientists in universities lose their advantage of having cheap labour because they still have to teach for some of the time and competitors have access to technical staff on a full-time basis. Competition for access to facilities in full-time laboratories will put them at a disadvantage simply because they are not 'on site' and close to the administrative system.

The growth of non-university, state-funded posts for public

science has weakened the links between university departments and reputational goals and boundaries as alternative
ways of organizing scientific work and priorities have become
established, especially in the 'directed' or 'finalized' fields.[47]
Even before this mushrooming of state laboratories, though,
the extent to which research in scientific fields was circumscribed by employment units in higher education was limited.
By no means all contributions in any field of x-ology came
from departments and institutes of x-ology and not all
scientists in such employment units limited their research to
such fields. This is partly because the competitive drive to
produce original contributions which lead to high collegiate
reputations encouraged specialization and differentiation so
that identities and boundaries entrenched in employment
units were no longer reflected in research strategies and
reputational communities. However, the pursuit of reputations for original contributions does not only encourage
differentiation within established boundaries but also leads to
attempts to establish new fields and areas of research as young
scientists try to acquire high reputations through developing
new methods and/or topics which do not fit in with current
identities and priorities. This is a high-risk strategy since it
threatens existing commitments and skills if resources, including prestige, for others are reduced. Hence, external support is
often required to carry it out successfully.[48] Established
university élites can control this process as long as they
control jobs and communication media, but where there are a
plurality of institutions providing these, and outside funding
agencies pursuing divergent objectives, their power is correspondingly limited.

Even within fields dominated by academic employment
organizations and university-based research facilities, there
are of course considerable variations in the degree to which
university élites form coherent and unified groups capable of
dominating research strategies and validation criteria. Professorial control of employment opportunities, access to
facilities and communication media in general in scientific
fields need not automatically imply that all professors are
united and act as a cohesive group. The greater the number of
employment units in a given field and the more easily

available are research resources, the more likely is it that mere incumbency of a chair will be of only limited importance in affecting the course of intellectual development. Influence over scientists' research strategies will have to be exercised more through the general reputational system than through direct control over resources and personnel at the work-place when alternative employment opportunities are available, and access to the necessary facilities and journals is easy. Authority in the reputational organization is here more important in controlling the direction and assessment of work than administrative authority in employment organizations. Generally, it seems reasonable to say that where the necessary resources of jobs, technicians, machines, and journals are widely available to most practitioners, and the control that any single professor, department, or institute has over them is small relative to the total pool available, administrative authority will be less important than reputational standing in affecting the course of scientific development. Professors, like other scientists, will have to compete in the wide reputational system for influence and authority if they wish to control scientific work. Control over resources is obviously a major factor in deciding the outcome of such competition but is rarely decisive on its own. Just as employment organizations are relatively independent of reputational organizations, and have their own patterns of development and change, then so too reputational organizations should not be regarded as simply tools of an all-powerful professorial power élite. Rather, variations in dependencies between these two sets of organizations need to be explored in different fields, together with their consequences for the structure of intellectual work.

EMPLOYEE-DOMINATED SCIENCES AS DUAL SYSTEMS OF WORK ORGANIZATION AND CONTROL

The institutionalization of scientific work and practices in the universities produced a hybrid system of work organization and control in which employers provided jobs, facilities, and trainees while work goals and performance evaluation were largely controlled by reputational groups which cut across employment unit boundaries. Universities certified scientific competence and employed the majority of knowledge producers, yet relied upon national and international networks to set work priorities and assess the correctness and significance of task outcomes. This hybrid system has been continued in state-supported laboratories which are primarily oriented towards the public science system, except that they are dependent on the universities providing certified skills. To a large extent this dual system of work organization and control has come to dominate the production and evaluation of public scientific knowledge as the major form of collective understanding in our society. Accordingly, the organization and development of that knowledge are dependent on the ways in which this system has functioned over the past century or so and different fields are organized differently according to variations in it.

The development of paid employment for researchers on a large scale, and the gradual decline of the unpaid 'amateur' as an important contributor to certified knowledge, have resulted in modern scientific knowledge being the product of employees. The organization and control of research by individual employers, and their relations with reputational groups, are therefore of considerable importance for the organization of knowledge. In particular the establishment of posts, institutes, and departments on a permanent basis obviously structures the sorts of subjects which will be studied and implicitly or explicitly organizes subjects into some sort of prestige ordering. Intellectual fields which are not entrenched in permanent positions will find it much more difficult to establish their intellectual and social autonomy than will those that have some control over posts and research facilities in employment organizations. Similarly, where state funds have become essential for continued production of knowledge which is generally regarded as a contribution to collective goals, individuals and groups who have only restricted or very limited access to such funds will obviously be disadvantaged. Reputational groups, then, who wish to continue to function as controllers of the production and validation of scientific knowledge in an era of employee-dominated science have to gain some control of employment and resource allocation decisions. Conceptions of science and of topics which do not manage to influence such decisions will, eventually, disappear from orthodox, establishment science.

It also follows from this point that those groups who do succeed in obtaining some control over posts and facilities will be in a strong position to impose their views of the correct approaches to be adopted and the relative importance of particular topics. Where senior positions are filled by proponents of a particular intellectual approach, or of a particular ordering of topics, obviously the direction of research in that field, and what is seen as constituting it, will be affected by their preferences as is illustrated by Foster's tenure of the post of Biological Secretary of the Royal Society from 1881 to 1903.[49] This will be especially so in fields where research can be planned in accordance with a strategy and laboratory directors are able to administer the work of a large number of people without requiring much consultation and direct collaboration between them as in twentieth-century chemistry.[50] However, the dual system of work control ensures that local control over scientific work is never complete and is mediatèd by reputational groups. Even where research is relatively routinized and can be administered by formal rules and structrues, and intellectual fields are dominated by a small number of scientists in directing positions, internal competition for reputations will ensure specialization and task differentiation so that sub-fields develop which may or may not be integrated into some overall hierarchy. Also, if reputations in the field are relatively socially and/or scientifically prestigious, outsiders may attempt to make substantial contributions, often by altering the criteria according to which knowledge claims are evaluated. This combination of internal differentiation and external permeability – which tendencies do of course vary between fields – reduce the likelihood of a small coherent group totally controlling jobs and reputations in the modern sciences.

Scientific fields vary, then, in the degree to which work is controlled by employers and reputational organizations and in the extent to which the boundaries and goals of these two sets of organizations overlap and coincide. Within the sytem of public science, i.e. where research is undertaken largely for publication and reputations and where employers share control over work goals and processes with international communication networks of scientists, there are considerable differences between the sciences in terms of direct employer control of work, the influence of national and international reputational networks, and the social and intellectual organization of such reputational systems. The more varied and contradictory are employers' goals, and the more divided and conflicting are research groups in reputational fields, the less likely are the latter to determine the goals and procedures used by most researchers in those fields. Instead, as in many biological fields, individual scientists have considerable autonomy from any single set of reputational leaders in deciding their research strategies and approaches.[51] Where, on the other hand, employers' goals and procedures are more oriented to reputational objectives – especially if they are focused on a single field – and the more coherent and integrated are reputational judgements and criteria, the more reputational networks will be able to control research goals and processes as well as controlling access to jobs and directing posts in employment organizations, as in much of twentieth-century academic physics.

These two dimensions – multiplicity of employers' goals and cohesion of reputational groups – interact and affect intellectual boundaries and development. For instance, where most employers are oriented to many goals which are divergent and incompatible it is unlikely that a single reputational organization will develop which is socially and intellectually coherent and integrated. Management studies appear to be an instance of this situation. Employment boundaries and goals, thus, affect those of scientific groups, especially where training programmes are similarly diverse and differentiated. The employment of skills and facilities for purposes and priorities in state science which are not entirely consonant with existing reputational goals has led to the formation of new reputational organizations and the weakening of existing ones' boundaries, particularly in the bio-medical sciences. Because employers tend to follow different strategies – both in terms of how they approach the same general goals and in terms of which skills they employ and how they organize these – these new fields rarely become very cohesive and strongly bounded. Where they rely upon skills and technical procedures acquired in other organizations, such as university departments, which are organized according to different principles, practitioners can orient their research to a number of audiences and so the intellectual and social boundaries of these new areas are highly permeable. The separation of employment from training and the diverse goals of employers here result in a plurality of reputational groups which share their control over work goals with employers and do not monopolize access to jobs or directing posts.

Even where employment organizations are primarily oriented to reputational goals, such as universities, there are considerable variations in the extent to which units of academic organization are identical to reputational boundaries and are thus likely to lead to relatively closed social and intellectual structures. Contributions to a particular field may come from scientists employed in variety of posts and employment units just as research in a given department or institute may be highly differentiated and contribute to a wide variety of intellectual goals. Generally, it seems probable that the more disjunction there is between the boundaries and goals of training units, employment units, and reputational organizations, the less likely the last are to become cohesive, oligarchic, and to control access to jobs and directing posts in employment organizations. The ability of any single reputational organization to monopolize labour markets and funding agencies will be limited in these circumstances and employment and promotion decisions will be the outcome of competitive struggles and negotiations between groups in different fields.

Where, on the other hand, the main units of employment organization are coterminous in their boundaries with those of training units and reputational groups, and employers are primarily oriented to reputational goals and performance criteria such as in twentieth-century mathematics, strongly bounded reputational organizations are more likely to develop and to control work goals. Directing groups in employment organizations are likely to be appointed largely on their international reputations and to allocate resources in accordance with reputational objectives and priorities. Having high reputational status and authority in employing organizations, such groups are in a powerful position to influence reputational goals and priorities and thus to determine the direction of research in these fields. This influence will be even stronger in fields where additional funds have to be obtained from a small number of national agencies to conduct research which contributes to reputational goals and these agencies are largely controlled by dominant reputational groups. Where, then, units of employment, funding, training, and reputational organization are largely the same in terms of goals and boundaries as in much of physics, they are likely to be dominated by the same group who will be able to set common criteria of performance assessment and insist on particular directions and procedures being followed. The more variations there are between the goals and boundaries of these units in employee-dominated research, the more diverse will scientific work tend to become. Differentiation of audiences, employers, and funding agencies reduces the pressures upon scientists to conform to a single set of objectives and ways of working and enables them to pursue divergent goals with novel techniques.

SUMMARY

The points made in this chapter can be summarized in the following way:

1. Before scientific research became dominated by employees it was relatively weakly demarcated from other systems of cultural production and was not highly organized into separate fields. Intellectual goals and reputations were more general and relatively open to individual influence while contributions to knowledge' were less specific and narrow than in the more 'professionalized' sciences. Research skills tended to be quite diffuse and non-standardized, they were often applicable to a wide range of problems and enabled considerable personal mobility between areas. Training in such skills was relatively unorganized and highly contingent upon personalities and

individual circumstances. Conflicts and controversies were quite intense and wide-ranging, and competing researchers sought to influence the direction of work across a number of areas and dominate the general reputational system. Intellectual commitments and standards were more personal and idiosyncratic than they later became; also control over research practices and goals tended to be personal and diffuse. Local variability was therefore quite considerable.

2. The growing dependence of jobs and income for an increased number of researchers upon intellectual reputations in the nineteenth century both extended and intensified reputational control of scientific work and integrated reputational goals and standards with employers' goals and authority structures. Thus reputations among specialist colleagues became more critical for scientists and more tied to particular arrangements for pursuing research in employment organizations. As employees began to dominate the system of knowledge production and evaluation, their organization and hierarchies became connected to reputational criteria and values.

3. Relations between employers' goals and reputational ideals can be summarized under three headings: *(a)* the degree to which employers' goals are oriented to public reputational ones, *(b)* the degree to which employers' personnel policies follow the verdicts and standards of reputational groups, and *(c)* the degree to which employers' rewards systems are dependent upon the standards and values of a single reputational élite in one field. Three types of science can be distinguished in terms of differences in these dimensions: industrial science, state science and academic science.

4. The particular institutionalization of knowledge production in the university systems of Europe in the nineteenth century, especially the German one, had four major sets of consequences for the organization and control of intellectual work. First, it integrated the production of intellectual innovation with their dissemination and the inculcation and certification of research skills. This encouraged the standardization of knowledge in the form of textbooks and

the separation of 'pure' research from non-academic work. Second, it led to the 'normalization' and specialization of research around hierarchically organized teams dominated by intellectual and administrative leaders. Third, it unified the production of skills with the production of knowledge and the organization of distinct labour markets in separate 'disciplines' which collectively constituted the system of legitimate knowledge. Fourth, it demonstrated how intellectual work could be organized and controlled and how knowledge producers could be systematically trained and certified.

5. The domination of knowledge production and validation by employees has resulted in a dual system of work organization becoming characteristic of the modern sciences in which control over how research is conducted and assessed is shared between employers and intellectual élites. Variations in how this control is shared across scientific fields affect patterns of intellectual organization and are important factors in understanding differences between the sciences.

Notes and references

[1] See S. F. Cannon, *Science in Culture*, New York: Science History Publications, 1978, chs. 4 and 5 for a discussion of 'pre-professional' science in nineteenth-century England and problems of constructing boundaries around 'physics'. For an account of how 'physics' changed in the eighteenth century, see J. L. Heilbron, 'Experimental Natural Philosophy' in G. S. Rousseau and Roy Porter (eds.), *The Ferment of Knowledge*, Cambridge University Press, 1980.

[2] Because the number of people involved in making scientific judgements was fairly small, and their interests were varied and changing, intellectual boundaries and opinions tended to fluctuate and be affected by individual influences more than when commitments and skills became entrenched in employment and training organizations.

[3] As reported by I., Grattan-Guinness 'Mathematical Physics in France, 1800–1835', in H. N. Jahnke and M. Otte (eds.), *Epistemological and Social Problems of the Sciences in the 19th Century*, Dordrecht: Reidel, 1981 at pp. 358–9.

[4] See for example, Roy Porter, *The Making of Geology*, Cambridge

University Press, 1977; 'Gentlemen and Geology: The Emergence of a Scientific Career, 1660–1920', *The Historical Journal*, 21 (1978), 809–36.

[5] Cannon, op. cit., 1978, note 1, pp. 39–40. According to Heilbron, op. cit., 1980, note 1, two-thirds of holders of 'physics' chairs in Germany in 1700 were medical doctors and 50 per cent of these left for careers in medicine. In 1800 the proportion was still a quarter.

[6] Cf. Porter, op.cit., 1978, note 4; J. G. O'Connor and A. J. Meadows, 'Specialisation and Professionalisation in British Geology', *Social Studies of Science*, 6 (1976), 77–89; M. Berman, 'Hegemony' and the Amateur Tradition in British Science', *Journal of Social History*, 8 (1975), 30–50; D. E. Allen, *The Naturalist in Britain*, London: Allen Lane, 1976; P. L. Farber, *The Emergence of Ornithology as a Scientific Discipline: 1760–1850*, Dordrecht: Reidel, 1982, pp. 125–40.

[7] H. Mehrtens, 'Mathematicians in Germany circa 1800', in H. N. Jahnke and M. Otte (eds.), *Epistemological and Social Problems of the Sciences in the Early 19th Century*, Dordrecht: Reidel, 1981 at p. 409.

[8] See, for instance, the papers by Grabiner, Scharlan, and Dauben on mathematics in the Jahnke and Otte (eds.) volume just cited. These authors seems to agree that the increased emphasis on the foundations of mathematics was connected to the need to teach the subject and develop a distinct identity.

[9] Especially those published by the various National and Royal Academies.

[10] Cannon, op. cit., 1978, note 1, p. 126.

[11] The influence of the 'Gentlemen of Science' in physics, astronomy, geology, mathematics, and geography as well as some other fields has been emphasized by J. Morrell and A. Thackray, *Gentlemen of Science*, Oxford University Press, 1981, pp. 26–9.

[12] As in, for example, the case of Liebig whose education is briefly described in J. B. Morrell, 'The Chemist Breeders: the research schools of Liebig and Thomas Thomson', *Ambix*, XIX, (1972), 8–9.

[13] Such disputes are not, of course unknown in the 'mature' professionalized sciences as recent studies of controversies indicate. See *Social Studies of Science*, volume 11, no 1, 1981 *passim*. For a historical case study, see R. G. A. Dolby, 'Debates over the Theory of Solution', *Historical Studies in the Physical Sciences*, 7 (1976), 297–404. On national differences in taxonomies, see Farber, op. cit., 1982, chs. 6 and 7.

[14] As Jungnickel says: 'Standing at the begining of the movement toward precision measurement in the natural sciences, Fechner had to uncover the principles underlying the procedures used arbitrarily by earlier observers and to develop new, perfected methods from those principles',: Christa Jungnickel, 'Teaching and Research in the Physical Sciences and Mathematics in Saxony, 1820–1850', *Historical Studies in the Physical Sciences*, 10 (1979), 36.

[15] See, for example, Karl Hufbauer, 'Social Support for Chemistry in Germany during the 18th Century: How and why did it change?' *Historical Studies in the Physical Sciences*, 3 (1971), 205–31; B. H. Gustin, *The Emergence of the German Chemical Profession, 1790–1867*, unpublished PhD thesis, University of Chicago, 1975, chs. 2 and 3; Maurice Crosland, 'Chemistry and the Chemical Revolution', in G. S. Rousseau and Roy Porter (eds.), *The Ferment of Knowledge*, Cambridge University Press, 1980.

[16] In addition to the papers cited in notes 1, 3, and 8, see H. J. M. Bos, 'Mathematics and Rational Mechanics', in G. S. Rousseau and Roy Porter (eds.), *The Ferment of Knowledge*, Cambridge University Press, 1980; R. H. Silliman, 'Fresnel and the Emergence of Physics as a Discipline' *Historical Studies in the Physical Sciences*, 5 (1974), 137–62. On the changing relations between astronomy and mathematics in an earlier period, see R. S. Westman, 'The Astronomer's Role in the Sixteenth Century', *History of Science*, 18 (1980), 105–74.

[17] Especially in England, cf. Cannon, op. cit., 1978, note 1; Morrell and Thackray, op. cit., 1981, chs. 1 and 5.

[18] As in, for example, nineteenth century debates in geology. Cf. C. Gillispie, *Genesis and Geology*, New York: Harper Torchbooks, 1959; R. Porter, op. cit., 1977. This feature, of course, is very much in evidence in the present human sciences and differentiates some of them from the more 'professionalized' sciences.

[19] Professionalization is used here primarily to refer to control over labour markets by skill training and certification agencies. Professional skills determine how work is conducted but not necessarily what it is done for.

[20] Increasing the number of competitors for jobs thus intensifies disputes over the significance of results and problems in a science, especially if alternatives are not available.

[21] Although many directors of R & D laboratories in industry have been successful public scientists.

[22] The importance of an industrial labour market for the expansion of modern physics is discussed by S. Weart, 'The Physics Business in America, 1919–1940,' in N. Reingold (eds.), *The Sciences in the American Context*, Washington D. C.: Smithsonian Institution, 1979.

[23] For recent discussions of this phenomenon see W.v.d. Daele *et al.*, 'The Political Direction of Scientific Development', in E. Mendelsohn *et al.* (eds.), *The Social Production of Scientific Knowledge*, Sociology of the Sciences Yearbook 1, Doedrecht: Reidel, 1977; R. Johnston and T. Jagtenberg, 'Goal Direction of Scientific Research', in W. Krohn *et al.* (eds.) *The Dynamics of Science and Technology*, Sociology of the Sciences Yearbook 2, Dordrecht: Reidel, 1978.

[24] N. Elias, 'Scientific Establishments', in N. Elias *et al.* (eds.), *Scientific Establishment and Hierarchies*, Sociology of the Sciences Yearbook 6, Dordrecht: Reidel, 1982.

[25] See, for example, Committee on research in the life sciences of the Committee on Science and Public Policy, *The Life Sciences*, Washington D. C.: National Academy of Sciences, 1970, pp. 230–9; E. Yoxen, 'Life as a Productive Force', in R. M. Young and L. Levidow (eds.), *Studies in the Labour Process*, London: CSE Books, 1981.

[26] As mentioned earlier, there is some dispute over the extent and significance of pre-nineteenth century employment of scientists. Roger Hahn and Maurice Crosland, for example clearly disagree over the existence of a career structure based on employment statuses in revolutionary and early empire France. See their papers in M. Crosland (ed.), *The Emergence of Science in Western Europe* London: Macmillan, 1975. See also. D. Outram, 'Politics and Vocation: French Science, 1793–1830', *British Jnl Hist. Science*, 13 (1980), 27–43. H. Gilmann McCann claims that French scientists had professional careers in chemistry between 1760 and 1795 in his *Chemistry Transformed: The Paradigmatic Shift from Phlogiston to Oxygen*, Norwood, New Jersey: Ablex, 1978. However, the numbers involved were very small and although high scientific reputations might lead to pensions and other forms of material rewards, there was no systematic machinery for linking jobs to research achievements throughout the sciences in the way which became established in the nineteenth century. On the Prussian reforms see C. E. McClelland, *State, Society and University in Germany 1700–1914*, Cambridge University Press, 1980; Roy Steven Turner, 'The Growth of Professorial Research in Prussia, 1818–1846, Causes and Context', *Historical Studies in the Physical Sciences*, 3, (1971); 'University Reformers and Professorial Scholarship in German', in L. Stone (ed.), *The University in Society*, Princeton University Press, 1974.

[27] As Everett Mendelsohn puts it: 'the university seemed to be the perfect home for science' in his 'The Emergence of Science as a Profession in Nineteenth Century Europe', in Karl Hill (ed.), *The Management of Scientists*, Boston: Beacon Press, 1964 at p. 43. Most accounts of professionalization in the sciences assume academic posts are the normal means of providing jobs for researchers. See the discussions in H. Kuklick,, 'The Organisation of Social Science in the United States', *American Quarterly*, 28 (1976), 124–41 and M. S. Larson, *The rise of professionalism*, University of California Press, 1977, chs. 1 and 2. On the 'professionalization' of chemists in nineteenth-century Britain outside the universities, see C. A. Russell *et. al.*, *Chemists by Profession: the Origins and rise of the Royal Institute of Chemistry*, Open University Press, 1977.

[28] This is emphasized by Turner, op. cit., 1971 and 1974, but McClelland suggests the change in the nineteenth century was not quite so marked. C. E. McClelland, op. cit., 1980, 342–3.

[29] McClelland, op. cit., 1980, note 26 chs. 3 and 4. Cf. Fritz Ringer, *The Decline of the German Mandarins*, Harvard University Press, 1969.

[30] Mendelsohn, op. cit., 1964 note 27; K. H. Manegold, 'Technology Academised: Education and Training of the Engineer in the 19th Century', in W. Krohn *et al.* (eds.), op. cit., 1978; W. Scharlan, 'The Origins of Pure

Mathematics', in H. N. Jahnke and M. Otte (eds.), *Epistemological and Social Problems of the Sciences in the Early 19th Century*, Dordrecht: Reidel, 1981. On the separation of university science from technology and 'practical research' in Sweden, see A. Jamison, *National Components of Scientific Knowledge*, University of Lund Science Policy Institute 1982, ch. 8.

[31] McClelland sees the particular form that the institutionalization of scholarship took in Germany, and its subsequent high social prestige, 'as an attempt to stabilize and legitimate the rule of the more flexible part of the aristocracy with the aid of a small élite recruited from the middle class', op. cit., 1980, p. 98.

[32] McClelland, op. cit., 1980, p. 106; Morrell, op. cit., 1972.

[33] As suggested by Terry Shinn, 'The French Science Faculty System, 1808–1914', *Historical Studies in the Physical Sciences*, 10 (1979).

[34] Porter, op. cit., 1978; Berman, op. cit., 1975.

[35] Morrell, op. cit., 1972; Gustin, op. cit., 1975, chs. 4 and 5.

[36] This process occurred later in the field sciences than in the laboratory sciences, and also later in England. See Allen, op. cit., 1976, note 6; O'Connor and Meadows, op. cit., 1976, note 6; Porter, op. cit., 1978, note 4; Farber, op. cit., 1982, note 6, ch. 7; Russell *et al.*, op. cit. 1977, note 27, ch. 9.

[37] Morrell, op. cit., 1972, note 12.

[38] Jungnickel, op. cit, 1979, note 14, pp. 41–2.

[39] See the discussions of Porter, Berman, and Cannon cited in notes 1, 4, and 6.

[40] Gustin emphasizes that Liebig was following an already established tradition in the training of apothecaries and so his innovativeness should not be over-emphasized, see Gustin, op. cit., 1975, note 15 ch. 3.

[41] Berman, op. cit., 1975; Jungnickel, op. cit., 1979.

[42] As discussed by Berman in his account of the Royal Institution in his *Social Change and Scientific Organization*, London: Heinemann, 1979. See also R. H. Wiebe, *The Search for Order, 1877–1920*, New York: Hill and Wang, 1968, ch. 6.

[43] R. Moseley, 'Tadpoles and Frogs: Some Aspects of the Professionalisation of British Physics, 1870–1939', *Social Studies of Science*, 7 (1977), 423–45; R. H. Kargon, *Science in Victorian Manchester*, University of Manchester Press, 1977, chs. 5 and 6.

[44] Heilbron, op. cit., 1980, note 1, points out that the rising costs of conducting research into electricity restricted the ability of casual observers to participate in research.

[45] Of course, raw materials in many natural history fields also required substantial resources to acquire as extensive travel became necessary and the 'Humboldtonian' science in general relied on numerous observers and state support. See, for instance, Cannon, op. cit., 1978 note 1 ch. 3; Farber, op. cit., 1982, note 6, chs. 3, 4, 5, and 6.

[46] T. S. Kuhn, 'The Function of Measurement in Modern Physical Sciences', in his collected essays *The Essential Tension*, Chicago University Press, 1977.

[47] G. Bohme *et al.*, 'Finalisation in Science', *Social Science Information*, 15 (1976), 207–330; R. Hohlfeld, 'Two Scientific Establishments which Shape the Pattern of Cancer Research in Germany', in N. Elias *et al.* (eds.), op. cit., 1982, note 24, 145–68.

[48] Direct state support was required in the German universities for chairs in new subjects and molecular biology received considerable support from the Rockefeller foundation just as genetics was helped by the Carnegie Institution. See McClelland, op. cit., 1980, pp. 174–89, 258–87; E. Yoxen, 'Giving Life a New Meaning: the Rise of the Molecular Biology Establishment', in N. Elias *et al.* (eds.), op. cit., 1982, note 24; Garland Allan, 'The Transformation of a Science: T. H. Morgan and the Emergence of a New American Biology', and Nathan Reingold, 'National Science Policy in a Private Foundation: The Carnegie Institution of Washington', in A. Oleson and J. Voss (eds.), *The Organisation of Knowledge in Modern America, 1860–1920*, John Hopkins University Press, 1979.

[49] See G. L. Geison, *Michael Foster and the Cambridge School of Physiology*, Princeton University Press, 1978, pp. 300–1.

[50] For mineral chemistry, see Terry Shinn, 'Scientific Disciplines and Organizational Specificity', in N. Elias *et al.* (eds.), op. cit., 1982, note 24.

[51] As characterized by B. Latour and S. Woolgar, *Laboratory Life*, London: Sage, 1979, ch. 3.

3

THE DEGREE OF MUTUAL DEPENDENCE BETWEEN SCIENTISTS AND THE ORGANIZATION OF SCIENTIFIC FIELDS

INTRODUCTION

IN the previous two chapters I have suggested that scientific fields are a particular kind of work organization which structure and control the production of intellectual novelty through competition for reputations from national and international audience for contributions to collective goals. As such, they deal with relatively high levels of uncertainty, compared to other work organizations, and rely on an extensive formal communication system to co-ordinate and interconnect task outcomes from a geographically dispersed set of work-places. The development of paid employment on a continuous basis for knowledge production in the nineteenth century, and the combination of general education, inculcation of research skills, and employment linked to reputations in many universities, gave the leaders of scientific fields greatly increased power over knowledge production at the same time as separating them into distinct disciplines controlling labour markets in particular skills. By connecting employment status and access to research facilities to national and international standing in a particular intellectual field, these developments organized knowledge production and validation around specialized interests and expertises. Thus, scientific fields become transformed into disciplines, that is, units of labour market control which trained knowledge producers in particular skills that monopolized contributions to particular intellectual goals.[1] Increasingly over the course of the nineteenth century, knowledge became identified with the products of these differentiated reputational organizations whose leaders combined control over skill definition and training, access to jobs and research facilities, and the dominant standards for assessing intellectual reputations.[2]

These academic disciplines incorporated and integrated three major components of intellectual production in the employee-dominated sciences: skill training and certification, access to material rewards and the facilities necessary for knowledge production, including academic posts, and the communication system. It is, though, important to remember that the degree of integration of these components varied between disciplines – such as between chemistry and ornithology[3] – and between historical periods. Since the extensive development of state-supported and 'steered' research, in particular, skill training has become more separated from direct control over knowledge production goals and employment in many of the public sciences.[4] Similarly, skills have become much more specialized and acquired in several distinct stages of training, some of which are conducted outside universities. Post-doctoral research, for example, has mushroomed in many fields since the 1920s in the USA and is now deemed critical to the development of élite scientists in the physical sciences.[5] Reputational networks are no longer identical to university employment unit boundaries in many scientific fields and particular research skills are now used for a variety of problems and topics in the pursuit of reputations which cross traditional disciplinary boundaries, especially in the bio-medical sciences. So changes in the context of research have affected the organization of scientific fields and these latter vary in their degree of integration and co-ordination of the different aspects of intellectual work in the modern sciences.

Scientific fields have become more important as distinct units of knowledge production and control, then, as they were entrenched in employment organizations and they exhibit different ways of organizing research which have altered over the past 150 years or so. They mediate between the broader social structures and the micro-processes of day-to-day research and negotiations over the meaning and significance of results. As such they constitute the major context within which public scientific research is carried out and made sense of. Variations in their patterns of organizing and controlling work through reputations, then, affect the scope of problems tackled, the extent of theoretical integration of results, the degree of competition between scientists and similar aspects of intellectual organization and development. Equally, these patterns of work organization and control are consonant with particular contextual circumstances and not others. Thus scientific fields with certain features are more likely to emerge and become established in some contexts than in others. General social processes in science as a whole, relations between private science and public science and between science and dominant social institutions all structure the sorts of scientific fields which develop and, through these, the sorts of knowledge structures which are produced and validated. Given the division of intellectual production into separate scientific fields, their entrenchment in employment and training organizations, and varying control over resources, their patterns of work organization and control are crucial features which structures the development and change of scientific knowledge. It is variations in these patterns, their consequences and contexts, which form the subject matter of the next four chapters.

DIMENSIONS OF WORK ORGANISATION AND CONTROL IN THE SCIENCES

The comparative analysis of scientific fields as reputational systems of organizing and controlling research requires some means of summarizing their most significant differences which are connected to different sorts of intellectual structures and to different sorts of environmental circumstances. While these could be drawn from the large literature on the administrative structures of work organizations and their causes and consequences,[6] reputational organizations have particular features which reduce the direct applicability of this literature, although I shall draw upon it when elaborating some details.

The most important of these features, of course, is the considerable degree of separation of control over how research is to be done, and competence assessed, from the immediate administrative hierarchy of employers. This renders much of the traditional literature on the formal means of planning and controlling work of limited utility, even in industrial science.[7] As a subset of professional systems of work organization and control, the employee-dominated sciences exercise some control over work processes and also over research goals in academic science. As we have already seen, this control can vary greatly between fields and affects the degree of intellectual integration and cohesion in them. University chemists, for instance, are able to control the nature and boundaries of research skills in chemistry to a much greater extent than can many academics in the social sciences and humanities where both lay groups and intellectuals from other fields are able to influence evaluation and competence standards. Similarly, élite physicists exercise more control over intellectual priorities and research strategies in many employment organizations than do many leaders of biological fields and the degree of theoretical integration and co-ordination of research in physics is much higher than in biology.[8] The extent of reputational control over employers' goals and use of research skills is, then, an important feature of scientific fields and the context in which they function.

Related contextual factors are the degree of independence or autonomy from lay concerns, languages, and concepts on the one hand, and from those of other, more prestigious, scientific fields on the other. The ability to maintain boundaries and distinct identities is a key feature of all work organizations which affects their degree of co-ordination of work and integration of goals. Where terms and procedures are similar to commonsense ones, or borrowed from other fields, it is obviously more difficult to separate problems and solutions from those of other groups and maintain unified control of research than in fields where vocabularies and work methods are more distinct and esoteric. The social sciences are an obvious contrast with the nineteen- and twentieth-century natural sciences in this respect.

Additional features of the environment of intellectual fields which can be distinguished here are the degree of concentration of control over access to resources and the diversity of audiences for intellectual products. Generally, the more limited are channels of access to the necessary means of intellectual production and distribution, the more dependent to scientists become upon the controllers of such channels and the more connected and competitive are their research strategies likely to be. Similarly, the more diverse and varied are legitimate audiences for their work, the more differentiated and separate are their goals likely to be so that co-ordination and integration of intellectual priorities in such fields will be limited. Thus the more open is a field to lay audiences and influences, the less probable are integrated theoretical structures which co-ordinate research throughout the field.

These environmental features of scientific fields, then, encourage particular patterns of work organization and control. These latter can be summarized in terms of two dimensions which stem from the nature of modern science as a system for co-ordinating and controlling the production of innovations as a collective enterprise. In characterizing employee-dominated scientific fields as a particular sub-type of professional organizations in Chapter 1, I suggested that their major distinguishing feature was the pursuit of reputations through producing novel contributions to collective intellectual goals. Essentially, modern sciences are systems of jointly controlled novelty production in which researchers have to make new contributions to knowledge in order to acquire reputations from particular groups of colleagues. These contributions are assessed in terms of their significance for collective goals and usefulness for others. They therefore have to be different and novel at the same time as being oriented to the work of colleagues and capable of being used by them in their own research. These conflicting demands create particular tensions between scientists and variations in their mutual balance affect the organization of knowledges that are produced in different fields. Many of the major differences between the sciences can derived from these variations which can be characterized in terms of two distinct dimensions: *the degree of mutual dependence between researchers* in making competent and significant contributions and *the degree of task uncertainty* in producing and evaluating knowledge claims.[9]

As collective systems of knowledge production and validation, the employee-dominated sciences have elaborate mechanisms for co-ordinating and ordering task outcomes and strategies such as journals, formalized communication systems, conferences etc. Through such mechanisms scientists structure the results of their work and attempt to persuade influential colleagues of its correctness and importance. In order to acquire high reputations for important contributions to intellectual goals, they have to demonstrate that particular results 'fit in' with those of others and are significant for their strategies. They are therefore quite dependent upon certain groups of colleagues who dominate reputational organizations and set standards of competence and significance. However, the extent to which scientists have to co-ordinate and specifically interrelate their research with that of a well-defined and bounded group of fellow specialists differs between sciences and over historical circumstances. In some fields, individual researchers are able to contribute to a number of distinct problem areas and seek reputations from different audiences by publishing results in different journals.[10] Elsewhere, for example particle physics, there is a clear hierarchy of journals and audiences are clearly defined. Contributions here have to be quite specific and to fit in with current concerns and results if they are to be acknowledged as useful.[11] The degree of mutual dependence between scientists varies, then, between fields and is related to differences in their intellectual structures.

As work organizations producing novelty and innovations, the sciences are also characterized by a relatively high degree of task uncertainty. The predictability, visibility, and repeatability of task outcomes is less in such organizations than elsewhere and sets limits to the development of formal planning and control systems. Again, though, the degree of task uncertainty varies between fields and affects perceptions of novelty and usefulness. Where there are a number of competing research schools, it is often not clear what the major issues in the field are, or how competence is to be assessed between such schools. Wundt suggested, for example, that a student of experimental psychology trained in the Leipzig laboratory would be bound to fail if confronted with examination questions from a representative of the Würzburg school.[12] In late nineteenth-century physiology technical standards were more widely agreed, but there were substantial conflicts over the significance of problems and the correctness of theoretical approaches between, say, Foster and his German competitors.[13] Some control over the extent of technical uncertainty has to be exerted, of course, for results from different work-places to be compared at all and reputations awarded on a national or international basis, but given that minimal limit it differs considerably between fields and such differences affect patterns of intellectual organization.

To some extent, dependence and uncertainty are necessarily interrelated in that the co-ordination of results across national boundaries implies sufficient control over phenomena for work to be communicated and compared. Equally, scientists need to be sufficiently dependent upon one another for reputations and rewards if they are to accept common evaluation standards and ways of doing research. However, within these constraints differing degrees of uncertainty can be associated with differing degrees of dependence so that scientific fields can be characterized in terms of variations on both dimensions. Furthermore, because we can examine the circumstances in which sciences change their positions on these variables, the analysis of the modern sciences in these terms provides a means of understanding scientific change as a social process. First, though, we need to discuss these two dimensions in more detail and see what factors are related to changes in them and what are their consequences for scientific research. Initially, I shall focus on the degree of mutual dependence, its implications for the organization and control of scientific research, and its relationship to particular contextual factors. Variations in task uncertainty will be the subject of the next chapter and both dimensions will brought together to characterize types of scientific fields in Chapters 5 and 6.

THE DEGREE OF MUTUAL DEPENDENCE BETWEEN SCIENTISTS

This dimension refers to scientists' dependence upon particular groups of colleagues to make competent contributions to collective intellectual goals and acquire prestigious reputations which lead to material rewards. Increasing the degree of mutual dependence implies that scientists become more reliant upon a particular group of colleagues for reputations and access to resources. They have to adhere to particular standards of competence and criteria of significance if they are to be awarded important reputations for their contributions. Thus they compete and co-operate with certain specialist colleagues and have to focus on a specific audience in seeking rewards. As mutual dependence in general increases, competition for reputations and control over the direction of research in that field grows in intensity, as does the strength of its organizational boundaries and identity.

The degree of mutual dependence has two analytically distinct aspects. The first is the extent to which researchers have to use the specific results, ideas, and procedures of fellow specialists in order to construct knowledge claims which are regarded as competent and useful contributions. This can be called *the degree of functional dependence* between members of a field and refers to the need to co-ordinate task outcomes and demonstrate adherence to common competence standards. Contributions which do not clearly fit in with existing knowledge and do not rely on similar techniques, methods, and materials as specialist colleagues are unlikely to be published in fields which exhibit a high degree of functional dependence. Furthermore, they must be demonstrably useful for others' research if they are to lead to high reputations in that field and so scientists need to show how their results could be incorporated in the work of colleagues when writing papers in highly dependent fields.[14]

The second aspect of mutual dependence refers to the extent to which researchers have to persuade colleagues of the significance and importance of their problem and approach to obtain a high reputation from them. This can be called *the degree of strategic dependence* for it covers the necessity of co-ordinating research strategies and convincing colleagues of the centrality of particular concerns to collective goals. Fields with a high degree of strategic dependence, such as German psychology where a few centres dominated knowledge production,[15] are highly competitive and involve considerable debate over the relative merits of different strategies and approaches to phenomena as scientists seek to direct one another's interests and strategies. Co-ordination thus is not just a technical matter of integrating specialist contributions to common goals but involves the organization of programmes and projects in terms of particular priorities and interests. It is a political activity which sets research agenda, determines the allocation of resources, and affects careers in reputational organizations and employment organizations.

These two aspects of mutual dependence between scientists are interconnected in that a high degree of one is unlikely to occur without a certain discernible amount of the other being present. A scientific field exhibiting considerable strategic dependence among its members, for instance, is likely also to manifest some functional dependence because researchers will need to demonstrate how their work fits in to, and is relevant for, the strategies of their colleagues if they are to convince them of its significance for general goals. Equally, high functional dependence implies some strategic dependence because scientists who need to use the specific results of colleagues and show how their own contributions are relevant are also claiming reputations on the basis of the importance of their strategies and approaches to their collective goals. Thus any reputational organization has to have a certain minimal level of both sorts of dependence if it is to function as a distinct system of work organization and control and a strong mutual orientation between specialists at one level implies some mutual dependence at the other. Nevertheless, within such limits some variation in the degree of functional and of strategic dependence does occur between the sciences which is related to variations in patterns of organization.

An example may illustrate this. If we compare modern physics and chemistry as distinct disciplines[16] it seems reasonable to suggest that both are highly specialized and exhibit a high degree of interconnectedness of procedures, topics, and results. They therefore can be described as having a high degree of functional dependence between their practitioners. However, whereas chemistry has become quite differentiated into separate sub-disciplines and specialisms which control resources and reputations in substantial autonomy from the discipline so that they do not appear to compete over their relative centrality, physics is much more hierarchically structured and has a distinct occupational group, theoreticians, for co-ordinating and ordering contributions within and between specialist groups.[17] Reputations in physics vary according to the prestige of one's sub-field and certain types of research are clearly considered more central than others. So physicists' strategic dependence upon one another and their need to demonstrate the importance of their area to physics as a whole is greater than appears in chemistry. Roughly similar degrees of functional dependence, then, occurs with different degrees of strategic dependence in contemporary scientific fields.

This example suggests that we can consider these two aspects of mutual dependence between scientists as varying relatively independently of each other within certain limits and so consider the sciences in terms of their position on each aspect separately. If we dichotomize these two dimensions and treat them as mutually independent, then four types of scientific field can be identified as in Table 3.1. Here, I have summarized a few of the major differences between fields with differing degrees of these two aspects of mutual dependence and suggesting some possible examples of each.[18] Additional differences between the sciences arise in connection with variations in task uncertainty, of course, and these will be discussed in the next chapter.

(a) In scientific fields where the degree of both sorts of mutual dependence between scientists is low, researchers are able to make contributions to a variety of goals without needing to incorporate specific results and ideas of particular specialist colleagues in a systematic way. They can deal with fairly broad problems and issues in a relatively diffuse manner and do not have to demonstrate exactly how their contributions fit in with those of other members of that field. Additionally, scientists are not greatly concerned here with persuading colleagues of the superiority of their approach to collective goals and do not seek to co-ordinate their strategies with others'. They therefore tend to pursue a variety of intellectual objectives with a variety of technical approaches and so contributions are rather diverse. Groups in these sorts of fields form around diffuse and general problem areas which are often characterized in commonsense and everyday terms

TABLE 3.1

Variations in the Degrees of Functional and Strategic Dependence in Scientific Fields

Degree of strategic dependence	Degree of functional dependence	
	Low	High
Low	(a) Weakly bounded groups pursuing a variety of goals with a variety of procedures. Little co-ordination of results or problems. Low extent of division of labour across research sites. Examples: Post-1960 Anglo-Saxon sociology, Management studies	(b) Specialist groups pursuing differentiated goals with specific, standardized procedures. Considerable co-ordination of results and specialized topics but little overall concern with hierarchy of goals. Examples: Twentieth-century chemistry, US mathematics
High	(c) Strongly bounded research schools pursuing distinct goals with separate procedures. High degree of co-ordination within schools but little between them. Strong competition for domination of field. Examples: German philosophy and psychology before 1933	(d) Specialist groups pursuing differentiated goals with specific, standardized procedures. Considerable co-ordination of results, problems, and goals and strong hierarchy of specialist concerns. Competition over centrality of subfields to discipline. Example: Twentieth-century physics

and are fairly fluid in membership and identities. There is considerable mobility between topics and areas and reputations fluctuate as interests alter. Competence standards tend to be relatively informal and diffuse with different groups interpreting them differently so that co-ordination of results between research sites relies heavily on personal contact and knowledge. These characteristics seem typical of many human sciences since the expansion of resources and jobs in the

1969s, especially perhaps Anglo-Saxon sociology and management research.[19]

(b) Where the degree of functional dependence is much higher but strategic dependence remains quite limited, scientists have to integrate their research much more with that of particular specialist colleagues and yet do not need to compete strongly over the relative merits of specialist goals for the whole field. They depend on the specific results of others' work to make competent contributions to collective concerns and so are more oriented to the research of particular groups than are scientists in the sort of field just described. This implies that competence standards are widely shared and fairly specific so that researchers are able to rely on colleagues' results with some confidence when making their own contributions. On the other hand, scientists in these fields do not have to integrate their specialist goals and compete over their relative importance in order to gain high reputations for access to rewards. So theoretical co-ordination of the various sub-fields is not very strong and the relative prestige of reputations in different specialisms is not the focus of intense conflicts. Reputations tend to be sought primarily in these sub-fields and problem areas rather than in the discipline as a whole, and integrative theoretical schemes which would order them in some system of intellectual priorities are not highly regarded or sought after. Examples of these sorts of fields are mathematics in the USA, artificial intelligence, many biological fields, and modern chemistry.[20]

(c) Where functional dependence between scientists is not so high but they are dependent upon one another for reputations based upon the significance and importance of their problems, strategies and results for overall goals, a rather different structure is apparent. In these fields researchers do not rely a great deal upon the specific work of particular specialist colleagues in making their contributions to collective goals but compete and conflict over the relevance of those contributions to such goals and the intellectual priorities of the field. The high degree of strategic dependence between practitioners means that scientists seek to persuade colleagues of the importance of their own approaches and topics by demonstrating the superiority of their interpretation of the central issues in the field. Battles in psychology over the crucial problems of the field, for example, also involved disputes over appropriate theoretical approaches and how the goals were to be achieved.[21] In these sorts of fields, then, the ranking of problems and approaches in terms of their importance to central goals implies a ranking of work procedures and competences as different research schools work on separate topics with distinct approaches so that results are often not easily compared across schools. Reputations tend to be awarded within such schools for contributions to their goals and the meaning of particular task outcomes for research done by other schools is difficult to discern or agree upon. Such specialization and division of labour as there is in such fields tends to occur within these general approaches and not between them. Rather, contributions are assessed in relatively diffuse and tacit ways with considerable reliance upon personal contacts and knowledge. Examples of these sorts of sciences occur mainly in the human sciences once they become dominated by academics, such as philosophy, early German psychology, literary studies, and perhaps British social anthropology in the 1950s and 1960s when resources were limited and research centres fairly tightly controlled by a few intellectual leaders.[22]

(d) The final type of scientific field to be considered in terms of variations of mutual dependence exhibits a high degree of specialization and division of labour coupled with a strong collective identity and consciousness of boundaries. Scientists are very concerned here to demonstrate how their work is important for others' goals as well as how it fits in with their research. The overall implications of task outcomes are critical for researchers if they wish to obtain high reputations and so they seek to convince disciplinary colleagues of the significance of their problems and concerns to the discipline as a whole. Competence standards are quite standardized and formalized throughout the field so that these implications can be drawn across specialist problem areas, and are strongly policed. Additionally, the importance of establishing the precise meaning of results for general goals and research throughout the discipline means that theoretical co-ordination of research strategies becomes a crucial activity. Research tends to be focused on relatively narrowly specified topics and conducted with standardized work procedures which generate particular, highly specific results. These are integrated through an elaborate communication system which uses a highly formalized symbol structure to co-ordinate research across geographical and social boundaries. Ambiguity is reduced by standardizing and formalizing reporting procedures so that the implications of task outcomes for general goals and strategies can be decided relatively easily and straightforwardly. The research system in such a discipline is therefore relatively centralized and quite formally co-ordinated. The obvious example of this type of field is modern physics.[23]

From these very brief remarks on different fields characterized by variations in the degrees of functional and strategic dependence between their members certain relationships between these dimensions and particular features of scientific fields can be discerned. Increases in the degree of functional dependence are associated with greater specialization of research topics and tasks, standardization of work procedures, competence standards and communication structure, and co-ordination of task outcomes from different research sites for dealing with particular problems. The scope of problems tackled by individuals and research groups tends to decline as functional dependence grows.

Increases in strategic dependence are associated with greater concern over the relative importance of problems and approaches, and so intensify competition between groups for domination of the field. Whether work procedures and competence criteria are standardized across groups or not, high strategic dependence implies a strong need to co-ordinate and interrelate research strategies and goals with those of specialist colleagues in order to gain important reputations from them. This, in turn, suggests a major concern with theoretical issues, similarities, and oppositions as competing groups explore the implications of their work and its meanings for rivals. Thus, the importance of theoretical co-ordination of contributions and approaches increases as strategic dependence grows. These, and other, relationships between variations on the degree of mutual dependence between scientists in a particular field and patterns of the organization and control of research will now be explored in greater detail, after which I will discuss some of the major contextual factors associated with changing degrees of mutual dependence.

VARIATIONS IN THE DEGREE OF MUTUAL DEPENDENCE AND THE ORGANIZATION OF SCIENTIFIC WORK

Changes in the degree of mutual dependence between scientists in a particular scientific field are associated with a number of substantial differences in patterns of work organization and control. Initially, I shall outline some of these which follow from increases in the general extent of mutual dependence. Secondly, I will discuss some which are particularly linked to increases in the degree of functional dependence, and lastly mention some changes which accompany increases in the degree of strategic dependence. The particular aspects of patterns of work organization and control which are associated with variations in the degree of mutual dependence concern the strength of intellectual identities and boundaries, the degree of specialization of tasks and skills, the degree and means of co-ordinating task outcomes and research strategies, the importance of general implications and theoretical significance of problems and results, and the scope and intensity of intellectual conflicts. Other aspects, such as the internal organization of scientific fields, the degree of intellectual cohesion and routinization of co-ordination procedures, are more related to variations in task uncertainty, or to the combination of these two dimensions, and so will be analysed in the following chapters.

(a) Increasing the degree of mutual dependence. The most obvious corollary of increasing dependence is a growing sense of collective self-consciousness and identity. If researchers become more dependent on a specific group of colleagues to obtain reputations which mediate access to material rewards, they will identify with that group more and separate themselves from other groups of knowledge producers and validators. Boundaries and distinctions between that repu-

tational system and its environment become firmer and more strongly policed as mutual dependence grows. Thus scientists develop sharp distinctions between competent researchers and outsiders, between relevant contributions and those not worth considering, between 'real' physics, economics, or whatever and trivial, uninteresting, and insignificant research. In Mary Douglas's terms they become 'high group' and exclude contributions from other fields.[24] Languages, work procedures, and cognitive objects become more esoteric and differentiated from other areas and from commonsense, everyday usages.

This separation and differentiation of knowledge production and evaluation systems has been most marked, of course, in the academically dominated disciplines. Here, a single organization controls the selection and training of recruits, the definition, inculcation, and certification of research skills, the allocation of jobs and access to research facilities, and the evaluation of intellectual products in terms of their correctness and contribution to disciplinary goals. Dependence upon disciplinary colleagues is obviously very high in such circumstances and contributions which transgress established norms and standards are not likely to be published, let alone lead to positive reputations. Problems and approaches which cross disciplinary boundaries will be ignored because they are most unlikely to enhance researchers' reputations among those who control access to critical resources. Thus the frequent calls for more inter- or multi-disciplinary research have achieved little unless accompanied by substantial resources and changes in the overall intellectual prestige of existing disciplinary reputations, as in the case of the invasion of biology by techniques and approaches from chemistry and physics.[25]

The growing strength of their collective identity among scientists who are increasingly dependent upon one another for reputations, is accompanied by the intensification of competition between them. As they become more oriented to a particular set of colleagues who are working on related problems, they develop greater awareness of this group as competitors for reputations as well as awarders of reputations in that field. The more circumscribed the field becomes, and the greater the exclusion of outsiders and their goals, standards, and procedures, the more the members of the field focus upon one another as collaborators in achieving collective goals and as competitors for making major contributions to those goals. This competition is not simply for recognition in a passive sense as discussed by Hagstrom and others[26] but involves an active, directing aspect as well. Scientists seeking high reputations from colleagues are not only seeking renown for their achievements in the organization; they also seek to direct others' research along particular lines and ensure that their interests, problems, and standards are accepted by colleagues in their own research. Because high reputations are granted for contributions which are used by others in their own work, and not just for general accomplishments, competition implies conflicts over the importance of problems, goals, and techniques and the direction of the field.

This intensification of competition between scientists in highly dependent fields is mitigated by the need to obtain the approval and support of colleagues. If they are to be convinced of the correctness and utility of one's work, they obviously have to share some of its presuppositions and objectives and so research which challenges the entire intellectual edifice of the field is unlikely to be undertaken in sciences where mutual dependence is high. Competition for positive reputations from a particular group of scientists implies an acceptance of their standards and procedures, at least to the extent of their being able and willing to use one's ideas and results. The more dependent a scientists is upon a particular group, the more he or she has to follow its norms if the research is to be considered competent, and so the intensity of both collaboration and competition grows with increasing mutual dependence. This suggests that the scope of disagreements declines although the importance of persuasion grows. The particular form that competition takes in a scientific field depends on whether functional or strategic dependence is greater and on the degree of task uncertainty.

A further aspect of increasing mutual dependence in a scientific field is the importance of national and international reputations in that field compared to local hierarchies and contingencies. Not only does it imply a sharper demarcation between scientific fields but it also reduces the impact and influence of purely local considerations in setting research strategies and evaluating performance. Thus, administrative directors who lack reputational standing in that field will be less able to affect what research is done, or how it is done, as mutual dependence among researchers increases. Equally, individual or employer idiosyncrasy becomes less tolerated as the professional group sets standards and work goals for all its members, and local autonomy in general is reduced as scientists have to convince particular specialist colleagues of their competence and the significance of their contributions. Intellectual variability and diversity based on access to local resources and interests declines, then, as the importance of particular reputational organizations grows. National and international leaders come to dominate strategy development, skill determination, and performance assessment as in the case of US physics in the 1920s.[27]

This increase in extra-local control of research obviously requires an efficient and extensive communication system for reporting results, co-ordinating task outcomes, and integrating strategies. Thus sciences could not have become established as reputational systems before the development of transport networks and some sort of postal system. More relevant here, though, is the development of a symbol system which can communicate task outcomes to a large audience without too much ambiguity. Furthermore, because this audience is often anonymous as far as the author is concerned, reliance upon tacit, shared personal experiences and understandings cannot be assumed. Thus, the language in which contributions are expressed needs to be quite specific and detailed, and yet impersonal and formally structured. Individual idiosyncrasy and personal interpretations are less important than the accurate reporting of generalizable procedures and outcomes in such a way that others can understand them and see their relevance for their own work. Largely local and individual circumstances and contingencies become irrelevant as dependence grows and it is the meaning of results for specialist colleagues in a wide variety of situations which is critical. Symbol systems become more formal and standarized under these conditions, then, and take on an air of abstraction and objectivity.

This, in turn, encourages research which can be effectively communicated through such a system.Work which requires a richer, more personal means of communication is less likely to be undertaken than research which can be reported in relatively formal and specific ways as scientists seek recognition from a largely personally unknown group who share some commitments and ideals through common training programmes but who may differ in their strategies, experiences, and approaches. The larger and more anonymous is the audience, and the greater the degree of mutual dependence, then, the more impersonal and formal becomes the symbol system and the more restricted and specific becomes the research to be communicated through it to gain high reputations. The increasingly technical nature of many fields in the humanities and formalization of the communication system in, say, philosophy and some approaches to historiography as they became 'professionalized' illustrate this process.[28]

(b) Increasing functional dependence.   In addition to these general features of the organization and control of scientific work associated with increases in the degree of mutual dependence between scientists in a particular field, there are some particular characteristics which develop with increasing degrees of functional dependence. These concern the standardization of skills and work procedures through common training programmes, the degree of skill and task specialization, and the growing specificity of task outcomes.

As scientists become more dependent upon a particular set of colleagues to make competent contributions to collective goals, they have to demonstrate their technical competence to those colleagues at the same time as producing original results. Innovations are, therefore, controlled and limited by the need to show how they fit in with others' work and standards. To do this they need to follow similar procedures and be sufficiently specific to be communicated across work sites without their meaning and utility becoming too ambiguous. If separate, individual task outcomes are to be co-ordinated for collective goals and lead to reputations based on their contribution to substantive intellectual problems, common technical backgrounds and methods have to be assumed. This is assured by the establishment of common training programmes throughout the field and the concomitant standardization of research skills and techniques. These also, of course, help to reinforce collective self-consciousness and limit access to the labour market in those skills. Increasing functional dependence, then, implies greater standardization of how research is conducted in different situations and of how research skills are taught and assessed. Examples of such standardizations of research procedures and tools are the development of common taxonomies and methods in the increasingly separate disciplines of natural history such as ornithology, botany, entomology etc. in the early nineteenth century,[29] and the analytical techniques of organic chemistry such as Liebig's apparatus for the combustion analysis of organic compounds which enabled him to create and dominate that field.[30]

Such standardization and enforcement of collective norms reduces the scope of legitimate contributions and thus of innovations. By insisting that particular colleagues have to be able to use results and ideas if they are to be regarded as worthy of recognition, these have to be tailored to that audience and match their interests and presuppositions. Task outcomes therefore become more specific and restricted in scope as scientists focus on producing results which fit in with the work of specialist colleagues. Rather than making diffuse and wide-ranging contributions which deal with substantive problems in the field, they will tend to use particular ideas and work of colleagues to construct novel but specific contributions which combine with others' to resolve particular problems in specific ways.

This restriction of the scope of topics tackled by individuals is connected to the growing specialization and differentiation of tasks. This, in turn, is encouraged by the increasing competition between members of scientific fields as mutual dependence grows. Such competition is limited in range because of the need to demonstrate fidelity to collective standards, but grows in intensity as collective dependence increases. Originality is therefore shown by differentiating topics and approaches to them while claiming legitimacy by following approved procedures. Rather than competing directly by working on identical problems, which is obviously a high-risk strategy, scientists in these fields prefer to tackle distinct but related issues which contribute to overall general goals without leading to sharp confrontations over priorities. As long as strategic dependence is limited, increasing functional dependence thus leads to greater specialization and differentiation of research topics and results, with a growing division of labour and tasks. An example of such specialization is contained in Edge and Mulkay's account of the development of radio astronomy in Britain where the Jodrell Bank and Cambridge groups pursued different topics with different instruments and different research styles.[31] The relative importance of these two approaches was not a critical issue as long as enough resources were available and they could divide up the field between them without having to compete over theoretical priorities to any great extent.[32]

(c) Increasing strategic dependence. It is precisely such competition over the relative importance and significance of individual and group strategies which grows with increasing strategic dependence. As researchers have to convince their specialist colleagues of the relevance of their goals and approaches to collective objectives more and more, they become much more concerned about what those objectives are, how they are to be interpreted and linked to research strategies, and how particular results are related to them. Growing strategic dependence involves increased debate over the implications and general merits of task outcomes and strategies so that theoretical co-ordination of research throughout the field becomes more important than when functional dependence alone is growing. For high reputations to be awarded in these circumstances, scientists have to persuade colleagues of the overall significance of their work for collective goals and so be able to trace its connections with other problems and approaches.

The importance of convincing a particular group of the centrality of one's contributions to their goals means that researchers cannot simply pursue separate strategies and topics without being much concerned about their connections and relevance for others' work. Instead they have to demonstrate the theoretical relevance of their strategies and how their problems contribute to the central issues of the field. They therefore need to be much more concerned with the theoretical connections and consequences of their topics and approaches than in fields where strategic dependence is low. This, in turn, involves a greater emphasis upon exactly how such connections are made and how the dominant goals are interpreted such that some strategies are accorded greater importance than others. In fields such as physics, for instance, the importance of tracing these connections has so enhanced the prestige of theoretical work that a distinct occupational group has become established which dominated the field.[33] This degree of theoretical elaboration and formalization is not, however, common because it implies the ordering of problem areas into a hierarchy of importance and centrality, as well as the downgrading of non-theoretical skills. Instead, scientists are more likely to pursue overtly theoretical objectives, be aware of the general implications of their research, and argue about its significance in terms of the field's goals.

In the absence of technical standardization and common working methods these arguments will not deal with exactly how results and ideas from different research sites fit together and are relevant for each other, but will tend to focus on alternative and contrasting approaches to central objectives which generate different sorts of results. Each local centre of research will follow particular strategies and interpretations of the field's goals with distinct procedures and outcomes which are not easily compared and co-ordinated. Different theories will be developed by different schools and linked to different ways of working so that debates over relevance and significance will focus on overall views about the field and its central problems. Psychology seems to be an example of this, at least before the Second World War.[34]

Where there is some technical standardization so that work can be compared across research sites and co-ordinated around common problems, high strategic dependence leads to greater concern with the particular implications of results from different groups for the general corpus of accepted knowledge. Innovations here have to fit in with and add to existing understandings if they are to lead to positive reputations. Competition between strategies involves demonstrating their specific consequences for others' work and their superiority for collective goals in co-ordinating research across problems. Schools still conflict over the relative merits of their approaches and topics but have to take account of how these deal with and incorporate other groups' results. They therefore differ in their intellectual priorities and perception of the central problems in the field but accept each others' competence standards and the correctness of many results. In physiology, Foster differed from his German competitors in his central problem and preferred theoretical orientation, for example, but followed standardized procedures and usually accepted their results.[35]

The importance of convincing colleagues of the significance of one's work where strategic dependence is high means that problems are more consciously selected with a view to their overall relevance and strategic implications. They therefore tend to be more general and theoretically derived than in fields whether the broad significance of topics is less critical. However, they are formulated with a particular audience in mind so that this generality is limited by their interests and goals. Claims about the importance and significance of a problem for the general of a field will, then, be restricted by the need to follow the currently dominant view of those goals. The degree of generality and theoretical innovativeness of problems is circumscribed by the concerns of the primary audience who will be reluctant to accept a subsidiary position in a general integrating scheme dealing with, say, the nature of 'life'. Theoretically oriented strategies and problems are encouraged in fields with a high degree of strategic dependence, then, but their scope and range is limited by researchers' dependence upon a particular specialist group.

In summary, differences in the degree of mutual dependence between scientists are associated with differences in the strength of boundaries and identities, in the degree of division of tasks and skills, in the scope of problems tackled by individuals and the specificity of task outcomes, in the degree of standardization of work procedures and skills, the intensity of competition between scientists, and the importance of general theoretical implications of results. Scientific fields exhibiting varying degrees of dependence between their adherents, then, vary in how research is organized, differentiated, and co-ordinated so that the sort of knowledge produced by individuals and groups differs in its diffuseness, generality, and formality. Variations in dependence are also associated with variations in the environments of scientific fields – including neighbouring reputational organizations – so that changes in certain key features of these environments encourage changes in the degree of mutual dependence between members of a field. These features, or contextual factors, will now be discussed.

CONTEXTUAL FACTORS AFFECTING THE DEGREE OF MUTUAL DEPENDENCE IN SCIENTIFIC FIELDS

Mutual dependence implies focusing upon the work and goals of a particular reputational organization. For this to be high, that organization has to be able to dominate the allocation of material resources and rewards so that scientists have to seek reputations from those specialist colleagues if they are to obtain access to those resources and rewards. This means that such reputations have high social and scientific prestige and alternative channels are not easily available. Thus three sets of contextual factors seem to be particularly important in affecting the degree of mutual dependence between scientists.

First, there is the general extent of autonomy and independence from competing organizations and the wider social structure. This includes the capacity to impose competence standards upon employers, to establish significance standards, and to develop a distinct language and set of descriptive terms and concepts. Second, there is the degree of concentration of control over access to resources. This includes the internal communication system and its relations to externally provided resources such as jobs, apparatus, technicians, field-workers, and funds. Third, there is the plurality and diversity of possible audiences for task outcomes and their relative status in the science system and elsewhere.

Relations between these three sets of factors and the degree of mutual dependence will now be discussed separately.

(a) Reputational autonomy. The ability to control skill and competence standards is a key feature of reputational organizations and the more knowledge of a particular set of problems or domain is produced only by possessors of particular skills, the greater is their mutual dependence. If employers can use a variety of skills, or organize them in ways which transgress existing reputational boundaries, to produce valid knowledge, then scientists' dependence upon a particular reputational group is limited. While this is the case in much industrial research, it also occurs in some state-supported and steered science where standardized skills are combined in novel ways for work on, say, medically defined problems. New reputational organizations have developed around some of these problems which reflect varied interests and influences so that performance standards are no longer totally controlled by the discipline which generated and certified the original research skills. Contrast, for example, reputational control over competence criteria in nuclear physics with that in many bio-medical research fields.

Control over significance standards is also a major factor in determining the degree of mutual dependence between scientists. Where the leaders of scientific field are able to determine the criteria of intellectual significance of contributions independently of employers' objectives, and those of other groups, scientists are obviously highly dependent upon them for reputations and seek to demonstrate the importance of their work in those terms. This situations is much more probable when the field is scientifically prestigious and able to control resources through the generally recognized value of its reputations. Where, on the other hand, significance standards are influenced by funding agencies pursuing a variety of goals, by other scientific groups of equal or greater prestige, and by employers, mutual dependence is much less and scientists can develop separate strategies based on divergent notions of intellectual importance. Again, this is apparent in many of the state sciences where objectives, resources, and jobs are not integrated into a common reputational organization, especially if their prestige is not very high. Contrast, for instance, plasma physics with epidemiological studies of cancer.

Control over the description of a field's domain and the admissable range of concepts and terms is a further critical factor affecting the degree of mutual dependence. The ability to exclude lay and everyday discourse from a competent contribution increases the importance of convincing fellow specialists of the merits of one's research and reduces the availability of non-professional audiences. This also applies to symbol systems from other fields since if knowledge claims can be legitimately made in terms taken from, say, chemistry in biological areas, the degree of dependence upon biologists is not very high. Mathematical forms of representation, of course, are effective in this exclusion function, although their use in the human sciences has had only limited success in augmenting reputational control because of their inability to impose coherent competence and significance standards upon employers and resources providers – except perhaps in analytical economics.

(b) Concentration of control over the means of intellectual production and distribution. In addition to reputational autonomy and ability to control standards being critical aspects of scientific fields which lead to greater mutual dependence, a further key factor is how that control is structured. Two aspects can be distinguished. First, there is the organization of the political system, especially control over access to communication media. Second, there are its connections to control over material resources and rewards to be considered. In both instances, it is the degree of centralization or concentration of control which seems to be the critical feature. The more oligarchic and centralized is control over crucial resources, the more dependent are scientists upon the leaders of that field.

The extent to which reputations are controlled by a relatively small élite varies considerably across scientific fields and is clearly related to the degree of mutual dependence between scientists. Contemporary Anglo-Saxon sociology and economics, for instance, differ greatly in the influence a relatively small number of intellectual leaders are able to exert over research startegies and significance criteria. They also differ greatly upon the extent of researchers' dependence upon one another for reputations and access to rewards, and the degree of theoretical pluralism.[36] Oligarchic control of reputations tends to be realized through control of the communication system which is obviously a key resource in a system premissed upon public reporting of results for the assessment of competence and performance. Liebig's journal, *Annalen der Chemie und Pharmacie*, was a critical resource in his domination of organic chemistry and Wundt's *Philosophische Studien* was equally key in his establishment of a distinct research school and 'knowledge factory' in psychology.[37] Generally, the more concentrated is control over the major communication media, the easier is it for a small élite to set standards and direct research strategies. Where each research group has access to a journal or press, and these are roughly equivalent in prestige, control over reputations is likely to be more equally distributed and researchers more able to pursue different strategies. In many fields, each research institute in Germany published its own journal or book series and this increased the degree of intellectual pluralism.

Control over communication media became much more influential when it was combined with control over other resources as scientists became employees who required expensive, collectively organized facilities. Indeed, if a journal could not provide reputations which lead to jobs etc. it seems unlikely that it would be much use as resource to influence research in the employee-dominated sciences. Publications which are not widely read and used have only limited effects: this is especially so in today's overcrowded communications systems. Reputational control increasingly involves control over access to the means of intellectual production as well as the means of dissemination.

Where, then, appointments to jobs and controlling positions in research groups are governed by standards set by a small group, these are also likely to govern the procedures and rules of communication media. According to Forman, Sommerfeld in Munich was the dominant influence in the allocation of physics chairs in Germany in the early 1920s and only a very limited number of other full professors were involved in such decisions.[38] Therefore, this élite group was able to set standards for the whole of German physics in this period, since these chairs dominated the production of knowledge in the major institutes. Because the dominant group was so small and controlled access to the critical resources, scientists were very dependent upon them and followed their standards and goals. Even though the major journal, *Zeitschrift für Physik,* was unreferred and unassessed, this élite was able to maintain control over publication standards through its domination of training programmes and jobs. Their priorities led to atomic physics dominating the discipline as a whole and becoming highly competitive in the 1920s.[39]

This oligarchic control of dominant posts in a field, and hence of reputational standards and careers, is reinforced by the use of very expensive equipment and buildings which are restricted in their availability and centrally controlled through the same standards. The high cost of building physics laboratories at the end of the nineteenth century meant that researchers had to depend on a few institute heads for access to the means of intellectual production and so further concentrated control in their hands.[40] The use of accelerators today is an equivalent phenomenon which adds to the centralization of control in physics and to the need to co-ordinate and rank proposals and problems. In general, it seems reasonable to say that any form of 'big science' is likely to encourage concentration of control in science and thus increase the degree of mutual dependence and co-ordination problems. Contrasting physics with chemistry, where most apparatus is available in university departments and access to jobs is not controlled by a small élite in most countries, we can see that the concern with theoretical unification and co-ordination is much less in the latter field than in the former and specialist sub-fields are not ordered into a single hierarchy of importance.[41]

In summary, the more control over access to journal space, jobs, apparatus, and funds is concentrated among a relatively small group of researchers who are fairly cohesive, the more they will dominate the reputational system and the greater is the degree of both functional and strategic dependence. As such concentration is reduced, so too the critical importance of discipline-wide reputations declines and strategic dependence between scientists and specialist sub-fields becomes less. If jobs and other resources are widely distributed among research sites, scientists will be able to develop their own strategies without having to convince a particular élite group of their significance in terms of a single set of theoretical standards in order to produce results. Thus, intellectual diversity is more probable where concentration of control is limited.

A related factor to the concentration of control over major resources in scientific fields is their size. Size, here means the number of knowledge claim producers who are competing for reputations in a given field. Generally, the larger this number is relative to available resources, the more intense will be competition for those resources and the more dependent scientists become upon one another for reputations leading to material rewards. Collins has suggested that this will lead them to specialize more and to standardize research procedures and focus on empirical topics as a way of universalizing evaluation criteria and mitigating competition.[42] Thus functional dependence increases with greater size, but not necessarily the degree of strategic dependence.

Generally, most studies of work organization have shown that increases in the number of employees have led to greater standardization of procedures, formalization of reporting structures and co-ordination mechanisms, and increased specialization of tasks,[43] and it certainly seems reasonable to assume that intensified competition will encourage scientists to differentiate their work by narrowing their concerns. However, the extent to which they will develop more standardized work procedures and formalized symbol systems depends upon the possibility and attractiveness of flight to other areas, the distribution of resources between research groups, and their internal political structures.

If the field has sufficient social and scientific prestige, and other fields have relatively abundant resources, then many researchers may react to intensified competition by changing fields, rather than co-ordinating their work more with present colleagues. Alternatively, if each research site controls sufficient resources to continue to produce and publish results, and if their internal structure is strongly hierarchical as in the traditional German university system so that junior scientists cannot easily obtain reputations leading to rewards by seeking the approval of colleagues in other centres, growth may simply lead to increased conflict over goals and procedures rather than standardized techniques and greater functional dependence. Where employers have less hierarchical structures, as in the élite US universities,[44] an increase in size probably has led to a narrowing of research topics and skills together with an insistence upon technical standardization to ensure that contributions are comparable and researchers can obtain positive, though not very high, reputations from the whole field by making specialized claims with reduced risks. But this need not necessarily be the case elsewhere, as the human sciences clearly demonstrate, and national variations in the organization of research employment are major influences here.[45] Large size, then, may encourage greater functional dependence between practitioners of a field but this relation is mediated by other factors.

(c) Audience plurality and diversity. The third general contextual factor to influence the degree of mutual dependence is the number and diversity of audiences for reputations. Where these are limited, similar in their goals and standards, and organized into a clear hierarchy of intellectual and social prestige, mutual dependence between scientists is high. Where, on the other hand, there are different audiences which are roughly equivalent in prestige and are oriented to distinct intellectual purposes, then the degree of dependence of scientists upon any single group is obviously much lower. This factor is also connected to the internal political structures of scientific fields because multiple and divergent environments decrease the effectiveness of highly centralized and oligarchic authority structures generally in work organizations,[46] and central control of research strategies is obviously more difficult when researchers can address divergent groups for reputations leading to rewards.

Audience plurality and diversity are encouraged by multiple resource-providing agencies pursuing different purposes and strategies and by the closeness of a field's concerns and concepts to everyday discourse and interest groups. The importance of medical goals, at least in legitimating resource claims, in cancer research, for instance, provides alternative foci and standards to purely scientific ones and so reduces the degree of mutual dependence between strategies and thus the coherence of the field.[47] Equally, the existence of the educated lay public as a legitimate audience in many of the human sciences has restricted their development of separate languages and standardized research procedures. The ability of intellectual leaders to dominate research strategies and insist upon particular standards of competence has been limited by the openness to everyday ideals, concepts, and descriptions in these fields, which has allowed rebellious groups to call on external resources and legitimation for support.[48] Additionally, the close links between some subjects and state agencies and social reform movements have similarly reduced the degree of mutual dependence between researchers in those fields and render the development of a high degree of technical and theoretical coherence rather unlikely.[49]

Similar points apply to scientific fields which are relatively low in general scientific prestige. Here, some scientists may seek to bolster their influence and reputations by adopting standards and approaches from more central fields, thus increasing the diversity of strategies and procedures. When such moves result in the general importation of techniques and skills from other fields, as in biology, the ability of existing disciplinary élites to continue to dominate strategies and competence standards becomes considerably reduced and so does the degree of mutual dependence. This point emphasizes the general importance of a field's location in the general science system for its degree of mutual dependence and the strength of its intellectual identity. The less central a field is to the currently dominant ideal of scientific knowledge, and the less prestige do its reputations have throughout the sciences, the more scientists are able and likely to seek alternative audiences and see less need to co-ordinate their work with that of specialist colleagues.

In conclusion then, the degree of mutual dependence in scientific fields is a product of the importance of their reputations, both in science and elsewhere, and the extent to which these monopolize access to critical resources. Sciences which have high prestige and autonomy from other social groups are more likely to exhibit a high degree of mutual dependence among their members, especially if resources are controlled by a relatively small élite. Where the environment is less controlled, researchers will be able to seek reputations from a variety of audiences and be less dependent upon a particular set of specialist colleagues.

SUMMARY

The points made in this chapter can be summarized in the following way:

1. Scientific fields have become important organizations mediating environmental influences upon research goals and producers. They are not necessarily identical to labour market units such as disciplines.
2. They can be understood in similar ways to other types of work organization, specifically in terms of their degree of mutual dependence and degree of task uncertainty.
3. The degree of mutual dependence between researchers in a field has two aspects: functional dependence and strategic dependence.
4. Changes in the extent of mutual dependence are associated with changes in the degree of collective self-consciousness, the intensity of competition, the degree of local and individual autonomy from collective goals and standards, and formality and restrictedness of the communication system.
5. Increases in functional dependence, additionally, occur concomitantly with growing standardization of skills and training programmes, specialization of tasks and procedures, and limitation of the scope of topics tackled by individuals and research groups.
6. Increases in strategic dependence, further, heighten concern with the co-ordination and mutual implications of research strategies and results. Theoretical issues become more critical and the focus of increasing attention. The

relative importance of problems and approaches also develops into a substantial source of debate.

7. Changes in the extent of mutual dependence are associated with changes in particular contextual factors. These are: the relative degree of reputational control over competence and significance standards and over descriptive concepts, the degree of concentration of control over access to the means of intellectual production and dissemination, and the plurality and diversity of audiences.

Notes and references

[1] Disciplines, therefore are seen as the institutionalization of scientific fields in training and employment units. They are primary skill inculcation and certification units which became linked to reputational organizations but are not necessarily identical to them. Thus scientific fields and scientific disciplines are not the same phenomena in my view. This distinction seems essential if we are to appreciate how the organization and control of scientific research has altered in the eighteenth, nineteenth and twentieth centuries. For an attempt to distinguish between disciplines and professions, see P. L. Farber, *The Emergence of Ornithology as a Scientific Discipline: 1760–1850*, Dordrecht: Reidel, 1982, pp. 114–32.

[2] For an account of how the 'gentlemen of science' restricted conceptions of knowledge and science in 'pre-professional' research, see J. Morrell and A. Thackray, *Gentlemen of Science*, Oxford University Press, 1981, ch. 5.

[3] As Farber suggests in passing: Farber, op. cit., 1982, note 1, p. 131. See also D. Allen, *The Naturalist in Britain*, London: Allen Lane, 1976, *passim* and S. F. Cannon, *Science in Culture*, New York: Science History Publications, 1978, chs. 4 and 5.

[4] On the modern 'steering' of science, see G. Böhme, W.v.d. Daele, and W. Krohn, 'Finalisation in science', *Scoial Science Information*, 15 (1976), 307–30; W.v.d. Daele, W. Krohn, and P. Weingart, 'The Social Direction of Scientific Development', in E. Mendelsohn, *et al.* (eds.), *The Social Production of Scientific Knowledge*, Sociology of the Sciences Yearbook 1, Dordrecht, Reidel, 1977; R. Johnston and T. Jagtenberg, 'Goal Direction of Scientific Research', in W. Krohn *et al.* (eds.), *The Dynamics of Science and Technology*, Sociology of the Sciences Yearbook 2, Dordrecht: Reidel, 1978.

[5] [R. B. Curtis], *The Invisible University, Postdoctoral Education in the United States*, Washington D. C.: National Academy of Sciences, 1969, chs. 1, 4. This is partly because of the narrowness and limited scope of PhD research in many fields and is much less noticeable in classical biology and the human sciences, see pp. 64–71.

[6] A useful summary and attempt at synthesis of this work is in H. Mintzberg, *The Structuring of Organisations*, Englewood Cliffs, N J: Prentice-Hall, 1979. See also D. S. Pugh, D. J. Hickson, and C. R. Hinings, 'An Empirical Taxonomy of Structures of Work Organisations', *Administrative Science Quarterly*, 14 (1969), 115–26. J. D. Thompson, *Organizations in Action*, New York: McGraw-Hill, 1967.

[7] On the limited utility of formal models of project selection and resource allocation in industrial research, see D. R. Augood, 'A Review of R and D Evaluation Methods', *IEEE Trans. Eng. Mgt.*, EM–20 (1973), 114–20; T. E. Clarke, 'Decision-Making on Technologically Based Organizations', *IEEE Trans. Eng. Mgt.* EM–21 (1974), 9–23; C. Freeman, *The Economics of Industrial Innovation*, London: F. Pinter, 1982, ch. 7.

[8] On the physics élite and dominant role of theorists, see J. Gaston, *Originality and Competition in Science*, Chicago University Press, 1973, pp. 26–31, 59–83; W. O. Hagstrom, The *Scientific Community*, New York: Basic Books, 1965, pp. 167–76, 247–52; D. J. Kevles, *The Physicists*, New York: Knopf, 1978, chs. 13, 22, and 23.

[9] For a similar approach see R. Collins, *Conflict Sociology*, New York: Academic Press, 1975, ch. 9.

[10] On the plasticity of fields, arenas, and papers in bio-medical research, see K. Knorr-Cetina, *The Manufacture of Knowledge*, Oxford: Pergamon, 1981, ch. 5; B. Latour and S. Woolgar, *Laboratory Life*, London: Sage, 1979, chs. 3, 4, and 5.

[11] On the importance of specific measurements in the physical sciences, see T. S. Kuhn, *The Essential Tension*, Chicago University Press, 1977, ch. 8. On the hierarchy of journals in particle physics, see J. Gaston, op. cit., 1973, note 8, p. 138.

[12] As discussed by M. Ash, 'Wilhelm Wundt and Oswald Külpe on the Institutional Status of Psychology', in W. D. Bringmann and R. D. Tweney (eds.,), *Wundt Studies*, Toronto: Hogrefe, 1980.

[13] As discussed by G. L. Geison, *Michael Foster and the Cambridge School of Physiology*, Princeton University Press, 1978, pp. 331–351. See also G. L. Geison, 'Scientific Change, Emerging Specialities and Research Schools', *History of Science*, 19 (1981), 20–40; G. Allen, *Life Science in the Twentieth Century*, Cambridge University Press, 1978, pp. 19, 82–94; D. Haraway, *Crystals, Fabrics and Fields*, Yale University Press, 1976, pp. 6–28.

[14] The importance of the 'results and discussion' section of papers and its focus on others' results and procedures demonstrates this point. See K. Knorr-Cetina, *The Manufacture of Knowledge*, pp. 121–6.

[15] Compare M. Ash, op. cit., 1980, note 12.

[16] Disciplines are taken to refer primarily to the main units of skill definition and labour market control. They are not necessarily identical to reputational organizations but constitute important constraints on the latter through their influence on jobs and appropriate competence standards.

[17] As discussed, for instance, by Hagstrom, op. cit., 1965, note 8, pp. 245–52. Thus physicists focus much more on theoretical issues than chemists who deal with relativey concrete and palpable properties of substances. See, for example, T. Shinn, 'Scientific Disciplines and Organizational Specificity', in N. Elias et al. (eds.), Scientific Establishments and Hierarchies, Sociology of the Sciences Yearbook 6, Dordrecht, Reidel, 1982. I should perhaps add here that although the level of strategic dependence in contemporary chemistry is lower than in physics, the common theoretical structure ensures that results are more integrated across specialisms than seems to be the case in, say, US mathematics. See, for instance, L. L. Hargens, Patterns of Scientific Research, Washington D. C.: American Sociological Association, 1975, ch. 2.

[18] These examples are meant to be illustrative rather than systematically worked through.

[19] See, for instance, N. Mullins, Theories and Theory Groups in Contemporary American Sociology, New York: Harper and Row, 1973 on the expansion of jobs in the 1960s and the emergence of five new theory groups. On management research, see R. D. Whitley, 'The Development of Management Studies as a Fragmented Adhocracy', Social Science Information, 23, 1984.

[20] Hagstrom, op. cit., 1965, note 8, pp. 228–35, characterizes this situation as anomic in mathematics. See also C. S. Fisher, 'Some Social Characteristics of Mathematicians and their Work', American Journal of Sociology, 78 (1973), 1094–118. On artificial intelligence, see J. Fleck, 'Development and Establishment in Artificial Intelligence', in N. Elias et al. (eds.), Scientific Establishments and Hierarchies, Sociology of the Sciences Yearbook 6, Dordrecht: Reidel, 1982.

[21] Perhaps more in Germany because of the limited number of job opportunities and domination of a few professors. See M. Ash, op. cit., 1980, note 11, and 'Academic Politics in the History of Science: Experimental Psychology in Germany, 1879–1941', Central European History, 14 (1981), 255–86. On the varieties of behaviourism in Anglo-Saxon psychology and their separation into distinct, opposed, schools see B. D. Mackenzie, Behaviourism and the Limits of Scientific Method, London: Routledge Kegan Paul, 1977, pp. 18–23.

[22] On British social anthropology, see A. Kuper, Anthropologists and Anthropology, The British School, 1922-1972, Harmondsworth: Penguin, 1975, ch. 5.

[23] The search for precision was not always a feature of what became physics in the nineteenth century. S. F. Cannon suggests it was more a characteristic of 'Humboldtonian' science; see Cannon op. cit., 1978, note 3, chs. 3 and 5. However, I wonder if it can be separated from the search for unity and theoretical coherence as physics emerged as a distinct intellectual field in the early to mid-nineteenth century. On changing conceptions of theories and experiments in research on electricity in the nineteenth century, see M. Heidelberger, 'Towards a Logical Reconstruction of

Revolutionary Change: the case of Ohm as an example', *Studies in the History of the Physical Sciences*, 11 (1980), 103–21.

[24] For a summary of Douglas's 'grid' and 'group' dimensions, see her 'Introduction to Grid/Group Analysis', in M. Douglas (ed.), *Essays in the Sociology of Perception*, London: Routledge Kegan Paul, 1982. It is not entirely clear how these two variables can be fully mutually independent of each other, nor what is the appropriate level of imputation, especially in highly differentiated societies. For varying interpretations see: D. Bloor, 'Polyhedra and the Abominations of Leviticus: Cognitive Styles in Mathematics', and M. Rudwick, 'Cognitive Styles in Geology' both in the Mary Douglas collection and K. Caneva, 'What Should We Do with the Monster? Electromagnetism and the Psychosociology of Knowledge', in E. Mendelsohn and Y. Elkana (eds.), *Science and Cultures*, Sociology of the Sciences Yearbook 5, Dordrecht: Reidel, 1981.

[25] As discussed by P. Abir-Am, 'The Discourse of Physical Power and Biological Knowledge in the 1930s', *Social Studies of Science*, 12 (1982), 341–82. See also E. Yoxen, 'Giving Life a New Meaning: the rise of the molecular biology establishment', in N. Elias *et al.* (eds.), *Scientific Establishments and Hierarchies*, Sociology of the Sciences Yearbook 6, Dordrecht: Reidel, 1982.

[26] Hagstrom, op. cit., 1965, note 8, ch. 1.

[27] Greatly assisted by philanthropic foundations and the development of post-doctoral fellowships. See, for instance, D. Kevles, *The Physicists*, New York: Knopf, 1977, chs. 13 and 14; S. Weart, 'The Physics Business in America, 1919–1940', in N. Reingold (ed.), *The Sciences in the American Context*, Washington D. C.: Smithsonian Institution. 1979.

[28] On the increasing restriction of philosophy to 'professional' concerns and domination of technically sophisticated areas, see B. Kuklick, *The Rise of American Philosophy*, Yale University Press, 1977, pp. 242–56, 451–80, 565–71. For a discussion of the segmentation and limited professionalization of the humanities in the USA before 1920, see, L. Veysey, 'The Plural Organised Worlds of the Humanities', in A. Oleson and J. Voss, *The Organization of Knowledge in Modern American, 1860–1920*, John Hopkins University Press, 1979.

[29] As documented by D. Allen, op. cit., 1976, note 3, and P. Farber, op. cit., 1982, note 1.

[30] As discussed in J. B. Morrell, 'The Chemist Breeders: the Research Schools of Liebig and Thomas Thomson', *Ambix*, 19 (1972), 1–46.

[31] M. J. Mulkay and D. O. Edge, 'Cognitive, Technical and Social Factors in the Growth of Radio Astronomy', in G. Lemaine *et al.* (eds.), *Perspectives on the Emergence of Scientific Disciplines*, The Hague: Mouton, 1976. D. O. Edge and M. J. Mulkay, *Astronomy Transformed*, New York: Wiley, 1976.

[32] Although see B. Martin, 'Radio Astronomy Revisited', *Sociological Review*, 26 (1978), 27–55 for a discussion of the competitive background to the development of British Radio Astronomy.

[33] Not without difficulty though. On the domination of experimentalists and their opposition to theoreticians in Germany, see P. Forman, *The Environment and Practice of Atomic Physics in Weimar Germany*, PhD dissertation, University of California at Berkeley, 1967, pp. 56, 69, 132–5; P. Forman, 'Alfred Landé and the Anomalous Zeeman Effect', *Hist. Stud. Phys. Sciences*, 2 (1970), 158, 217.

[34] See the work of Mitchell Ash, op. cit., 1980, note 12; 1981, note 21 on German psychology.

[35] As discussed by Geison, op. cit., 1978, note 13.

[36] Compare, for example, Mullins, op. cit., 1973, note 19; N. Wiley, 'The Rise and Fall of Dominating Theories in American Sociology', in W. E. Snizek *et al.* (eds.), *Contemporary Issues in Theory and Research*, London: Aldwych, 1979; H. Katouzian, *Ideology and Method in Economics*, London: Macmillan, 1980, ch. 5; J. R. Stanfield, *Economic Thought and Social Change*, Carbondale, Illinois: Southern Illinois University Press, 1979, ch. 8.

[37] On Liebig, see, Morrell, op. cit., 1972, note 30; B. Gustin, *The Emergence of the German Chemical Profession, 1790–1867*, unpublished PhD dissertation, University of Chicago, 1975, pp. 92–102; on Wundt, see Ash, op. cit., 1980, note 12.

[38] Forman, op. cit., 1967, note 33, p. 107.

[39] Ibid., pp. 178–95, 316.

[40] P. Forman, J. Heilbron, and S. Weart, *Personnel, Funding and Productivity in Physics circa 1900*, Historical Studies in the Physical Sciences 5, Princeton University Press, 1975, pp. 85–114.

[41] The NAS Chemistry Survey Committee saw the field as 'little science' in its *Chemistry: Opportunities and Needs*, Washington D. C. National Academy of Sciences, 1965. See, also, S. S. Blume and R. Sinclair, *Research Environment and Performance in British Chemistry*, London: HMSO, Science Policy Studies, no. 6, 1973, Table II.7.

[42] R. Collins, op. cit., 1975, note 9, pp. 510–11.

[43] Especially the 'Aston' studies. See, for example, D. S. Pugh and D. Hickson. *Organizational Structure in its Context*, Farnborough: Saxon House, 1976. Compare, W. H. Starbuck 'Organizational Growth and Development', in J. G. March (ed.), *Handbook of Organizations*, Chicago: Rand McNally, 1965.

[44] As discussed, for example, in J. M. Beyer and T. M. Lodahl, 'A Comparative Study of Patterns of Influence in United States and English Universities', *Administrative Science Quarterly*, 21 (1976), 104–29.

[45] For a general discussion of national intellectual styles see J. Galtung, 'Structure, Culture and Intellectual Style: an essay comparing saxonic, teutonic, gallic and nipponic approaches', *Social Science Information*, 20 (1981), 817–56.

[46] As Thompson suggests, see J. D. Thompson, op. cit., 1967, note 6, p. 129.

[47] The conflict between medical and scientific establishments in cancer research is described in R. Hohlfeld, 'Two Scientific Establishments which Shape the Pattern of Cancer Research in Germany: Basic Science and Medicine', in N. Elias *et al.* (eds.), *Scientific Establishments and Hiearchies*, Sociology of the Sciences Yearbook 6, Dordrecht: Reidel, 1982.

[48] Most notably perhaps in sociology but also in literary studies where lay readers and the literary public are involved in disputes over goals and methods of work.

[49] Philip Abrams has outlined this process in British sociology in his *The Origins of British Sociology: 1834–1914*, University of Chicago Press, 1968, See, also, H. Kuklick, 'Boundary Maintenance in American Sociology: Limitations to Academic "Professionalisation" ', *Journal of the History of Behavioral Sciences*, 16 (1980), 201–19.

4

THE DEGREE OF TASK UNCERTAINTY AND THE ORGANIZATION OF SCIENTIFIC FIELDS

TASK UNCERTAINTY IN THE SCIENCES

THE modern sciences are institutionally committed to the production of novelty and only contributions which are recognized as new can lead to high reputations and other rewards in the public sciences. Research activities are therefore quite uncertain compared to other work activities, in the sense that task outcomes are not repetitious and highly predictable. Each result is unique and different from all others in at least one important respect so that work procedures cannot be fully pre-planned and routinized. However, novelty is recognized as such in terms of some background expectations and assumptions. Without some common knowledge and views about what should occur in a given research project it is unclear whether, and to what extent, results are novel and important. The production and recognition of new knowledge depends, then, on the existence and structure of current knowledge and expectations. The more systematic, general, and precise is existing knowledge, the clearer will any results be in terms of their novelty and significance for this common stock of understanding. Task outcomes in scientific fields where such background knowledge is widely shared and organized into a relatively coherent system will tend to be more predictable, and their implications more easily drawn, than they are in fields where the amount and coherence of existing knowledge which is shared between practitioners is not so great. In Kuhnian terms, the more paradigm-bound a field is, the more predictable, visible, and replicable are research results and the more limited is permissible novelty.

Variations in the extent to which work procedures, problem definitions, and theoretical goals are shared between practitioners, and are clearly articulated, are thus related to the degree of task uncertainty in scientific fields. Although results are novel in all sciences, their degree of novelty varies and so does the ease with which they can be fitted into existing knowledge and their implications drawn. Task uncertainty, then, differs between scientific fields and these differences can be related to other dimensions of variation between the sciences. Just as the degree of routinization of tasks and technological processes in other work organizations has been seen as a major explanatory factor in accounting for differences in organizations' structures,[1] so too we can expect to find considerable differences between scientific fields which vary in the degree to which task outcomes and research processes are predictable, visible, and clearly related to general goals. In general, routinization of tasks seems to be associated with standardization and formalization of control procedures, centralization of authority, a high degree of division of labour, and co-ordination of outcomes through pre-planned procedures. The discretion and influence of workers are highly restricted in organizations where technologies and tasks are relatively routinized, according to most of the literature in this field.[2] Before, though, proceeding to analyse these relations in more depth it is important to clarify the notion of task uncertainty in a little more detail.

There are a number of different ways in which task uncertainty and variability have been discussed in the organizational literature, which are not all mutually compatible. Some common elements, though, are shared by most writers on technology and organizations. These include the degree of uniformity and stability of raw materials, the degree to which materials and procedures are well understood and standardized, the variability of goals and desired outcomes, the rate of change of goals and techniques, and the number of exceptions which people have to deal with.[3] Most authors end up by constructing various taxonomic schemes with these or similar dimensions and suggest how different sorts of organizations can be allocated to one of the boxes and exhibit particular structural properities as a consequence.[4] For instance, Perrow has analysed the degree of routine in terms of the existence of well-established techniques which can be reliably assumed to produce the right results and the uniformity and stability of raw materials and tasks.[5] He suggests that work organizations can be analysed in terms of these mutually independent variables which have substantial consequences for systems of work control and co-ordination. This sort of distinction between the variety of problems and tasks, including their rate of change, and the straightforwardness of technical procedures which should be applied to them is fairly common in the organizational literature and can be extended to the sciences.

Scientific research is often thought of as being highly methodical and systematic in its technical procedures so that results are stable and replicable. Research methods and techniques are so standardized that they can be taught by textbooks and formal training programmes as well as being entrenched in elaborate machniery and apparatus. However, the tacit component of research has, of course, been emphasized by many writers such as Polyani and Ravetz, and recent studies of the research process have highlighted the highly contingent and *ad hoc* nature of much empirical research.[6] Furthermore, any social scientist knows that methods as specified in textbooks and courses are rarely directly applicable and in general the degree of ambiguity of technical procedures and of the meanings of their results can be quite high in the sciences and varies between them.

The extent to which work techniques are well understood and produce reliable results in various scientific fields can be termed *technical task uncertainty*. In fields where this is relatively high results will be ambiguous and subject to a variety of conflicting interpretations and the use of technical procedures will be highly tacit, personal, and fluid. It will not be very obvious when particular methods should be used, nor when they have been applied successfully. Where technical task uncertainty is lower, on the other hand, there will be a well-established set of research techniques which can be acquired through formal training programmes and whose use is relatively straightforward and success is easy to determine. Research results will be more predictable, visible, and replicable in these fields than in others. Given a particular problem, scientists will be fairly clear about how it should be dealt with, how the appropriate methods should be applied, and how the results should be made sense of, whereas in fields with higher technical task uncertainty none of these points will be easy to determine and disputes over them may be common.

The second dimension suggested by Perrow – the variety of problems and tasks to be performed – refers both to the degree of standardization of raw materials and to the variability of objectives. In the sciences the development of 'pure' elements and homogeneous, stable phenomena was a key point in the emergence of modern chemistry and other natural sciences because it enabled technical control of the subject matter to be widely demonstrated and greatly increased, as Rip points out.[7] However, I think that this aspect of scientific work is strongly related to the establishment of technical task certainty in a field in that it is difficult to see how work procedures can be well established and understood without some degree of control over the phenomena being studied. Standardization of techniques implies standardization of raw materials to the extent that objects are constructed in such a way that general procedures can be applied to them. To use a particular technique which has been standardized, scientists have to develop descriptions of the object being analysed which are commensurate with the procedure and this requires some restriction of its properties and uniformity of features. Indeed, it is difficult to see how reliable and well-understood techniques could be used in a relatively routine manner without the objects to which they are applied being similarly well understood and highly controlled. Rather, then, viewing this second dimension as focusing on the actual cognitive objects being worked on such as chemical elements or biological molecules, it seems better to consider uniformity and stability as dealing with the nature of the problems being considered and so analyse scientific fields in terms of the variability and fluidity of the problems posed and priorities established.

In discussions of work organizations this dimension is often taken to refer to the variability of products and/or markets and occasionally generalized to the overall level of environmental uncertainty.[8] Given the institutional commitment to innovation and differentiation of cognitive products in the sciences, this interpretation does not seem to be directly applicable because product variability is built into the research system. However, sciences do vary in their willingness to consider problems of widely differing natures and in their toleration of alternative approaches to the dominant one. Equally, the stability of problem formulations, and of hierarchies of problems according to their importance and significance, varies across fields so that, suitably modified, this dimension may be used to differentiate sciences and suggest reasons for variations in knowledge structures and patterns of change. It can be called *the degree of strategic task uncertainty*, since it encompasses uncertainty about intellectual priorities, the significance of research topics and preferred ways of tackling them, the likely reputational pay-off of different research strategies, and the relevance of task outcomes for collective intellectual goals.

Scientific fields exhibiting a high degree of uncertainty in this sense deal with a large number of different sorts of problems whose formulations quite legitimately differ and whose importance is subject to alternative evaluations which may be fluid and rapidly changing. The variety of problems is thus considerable, differences about how they should be best conceived are also evident and these conceptions are relatively unstable. Additionally, the 'value' of these problems to possible publics is indeterminate and shifting. The variability and stability of problems in scientific fields, that is, involves general considerations of intellectual goals and ideals in those fields, and in the sciences as a whole. The less clear it is which problems are most important, and the more those evaluations are subject to rapid change and local influence, the greater the variability of problems dealt with in different research sites and thus in the field as a whole. Uncertainty about appropriate goals is therefore a key aspect of this dimension. The degree to which there is a clear ordering of goals and problems which is generally accepted by practitioners in a given field directly affects the uniformity and stability of research activities. Where there are a number of different groups pursuing different goals to whom particular results might be of interest, scientists have a variety of possibilities in deciding what to work on and how to go about it; thus goals

TABLE 4.1

Variations in the Degree of Technical and Strategic Task Uncertainty in Scientific Fields

Degree of strategic task uncertainty	Degree of technical task uncertainty	
	Low	High
Low	Considerable predictability, stability and visibility of task outcomes. Implications of results easy to draw and relatively uncontroversial. Problems and goals fairly clearly ordered, restricted and stable. Examples: twentieth-century chemistry; physics since 1930s	Limited technical control of empirical phenomena, results unstable and difficult to interpret. Implications of task outcomes subject to alternative views and difficult to co-ordinate. Problems and goals restricted, stable, and tightly structured. Example: Economics since 1870.
High	Considerable predictability, stability, and visibility of task outcomes. Consensus on how to interpret results and co-ordinate them. Problems and goals rather varied, unstable, and not clearly ordered. Examples: Biology since 1950; artifical intelligence; engineering; pre-darwinian ornithology.	Limited technical control of phenomena, results unstable and difficult to interpret. Varied views about the implications of task outcomes and little co-ordination and comparison of them. Problems and goals also varied, unstable, and conflicting Examples: Post-1960 US sociology; post-1960 US ecology

and problems are much more varied – and liable be changed – than they are in situations where it is generally accepted what the most important problems are, what sorts of solutions are admissible, and which group is able to decide on the truth status of knowledge claims.

In the literature on work organizations these two dimensions are commonly viewed as being mutually independent and organizations classified in terms of their positions on each in simple taxonomic tables.[9] Scientific fields can also be distinguished in terms of these two types of uncertainty considered independently as in Table 4.1 although certain combinations seem more likely to occur than do others. For example, the combination of low strategic task uncertainty with high technical task uncertainty is not likely to be very stable because technical variability and instability means that results are difficult to compare systematically and so their implications for theoretical problems are unclear and subject to different interpretations. In this situation scientists are likely to develop locally idiosyncratic ways of dealing with technical difficulties which are highly tacit and dependent upon personal knowledge for transferrability. In other types of work technical uncertainty is feasible with product uniformity because craft procedures reduce uncertainty sufficiently for uniform task outcomes to be produced reliably – as in the case of the construction industry and the glass industry.[10] In the sciences, though, high technical uncertainty makes it difficult to see when products are similar or widely varied because task outcomes are non-standard and subject to various interpretations and descriptions. Where, in other words, working methods are not clearly understood and do not produce visible, predictable, and replicable results the uniformity and stability of their products cannot be vary high because it is impossible to be sure what they are in any standard way. Instead solutions to problems will vary considerably across work-places, and so too will the ways in which problems are understood so that standardized, stable formulations are improbable.

Restriction of the scope and variety of problems is perhaps possible in some scientific fields where technical task uncertainty is high, however, by effectively decoupling empirical research from theoretical issues and insisting on the priority of theoretical uniformity and stability. Admissible problems in such fields are highly restricted in type and conception; deviant formulations are likely to be ignored or relegated to 'applications' or simply regarded as indications of professional incompetence. This restriction of problems and concepts allows some research techniques to be quite standardized and formal but the general level of uncertaintly about task outcomes means that interpretations of empirical results become difficult to organize into a coherent scheme which bears directly upon the theory considered. Technical task uncertainty, that is, remains sufficiently high in such fields that research outputs can be described in various ways despite common technical procedures being used. This is because the raw materials are not sufficiently closed and restricted for their behaviour to be tightly controlled. In economics, for example, the highly formal techniques used in econometric analysis still require substantial interpretative skills to be applied and some of the largest variances in forecasts are due to 'judgemental' differences between groups.[11] Technical control of empirical phenomena in such fields is obviously low compared to that current in chemistry and some biological sciences. Rather, though, than this uncertainty about task outcomes leading to variability and instability of problems and theoretical approaches, as it has done in some other social sciences, in economics it seems to have encouraged a separation of formal theoretical work from empirical analyses of macro-economic data and a heightening of the prestige of the former. The intractability of the data has lead to the most central parts of economics becoming insulated from empirical problems and achieving intellectual certainty by becoming more theoretical and mathematical. As Phyllis Deane puts it: 'None of these doubts, however, was allowed to threaten the analytical core of "pure economics" . . . whatever the intractability of the problems posed in the applied areas of their discipline, they were building on virtually impregnable analytical foundations.' Restrictedness has been imported from mathematics, to use Rip's terminology, but not as a means of exerting technical control over empirical objects.[13]

In general, few fields seem to have followed the example of economics in this direction although periodic attempts to formalize the social sciences have been made. Instead, I think that most scientific fields can be characterized as being either high in terms of both forms of uncertainty, or exhibiting relative technical uncertainty but a high degree of problem variability and instability or a low extent of both forms of uncertainty. These three types can be construed as representing a continuum of decreasing uncertainty in the conduct, co-ordination, and control of research in the modern sciences. Their characteristics will now be briefly discussed before examining in more detail the major factors associated with changes in both types of task uncertainty.

Fields with the highest degree of task uncertainty are perhaps found most in the humanities and social sciences. Here, craft techniques are held in common – though to varying degrees – sufficiently for task outcomes to be viewed as being related but results are relatively unpredictable, subject to a variety of different interpretations and cannot be reproduced with a high degree of reliability. Theoretical goals are also disputed and are only weakly connected to technical procedures and task outcomes so that the significance and relevance of results for particular theoretical goals are difficult to establish. The range of problems which are relevant to practitioners is large and boundaries are unclear and also subject to disagreement. Additionally, the importance or problems is uncertain and different groups assess them in opposing ways together with appropriate theoretical strategies. The stability of problem formulations and of audiences is also low so that scientists do not know with any degree of certainty how the major issues should be conceived or which group should be addressed. This gives individuals a considerable degree of autonomy from any single dominant group in formulating his or her research strategy but also means they have little information for deciding on the 'best' way to generate valuable results which will lead to high reputations from important groups. Depending on the extent to which they have to co-ordinate their results with the work of others, scientists in these fields engage in constant negotiations and conflicts over priorities and methods. The intellectual structures are highly fluid and unstable with relatively little agreement on what the appropriate problems are, or the proper approaches and techniques. Audiences are diverse, rapidly changing, and adopt various evaluation criteria.

The second type of scientific field exhibits lower technical task uncertainty but relatively high strategic uncertainty. Results are here fairly stable, produced with standardized techniques, and rely on standardized substances or phenomena. The predictability, visibility, and reliability of task outcomes is quite high so that tasks can be differentiated between skills and their results co-ordinated around general goals. However, problems cover a side range of issues and can be conceived in a variety of ways so that scientists can pursue diverse goals and theoretical divergencies are considerable. The significance of results for overall objectives is difficult to establish with any degree of stability in this situation and different groups will have different evaluations of the relevance of task outcomes. Practitioners in these fields have a common body of fairly standard skills which are systematically inculcated in training programmes, and which generate definite and comparable results across research sites, but disagree about the general relevance and significance of research strategies for organizational purposes. Indeed, they may pursue a variety of different objectives and audiences with no clear theoretical strategy in mind and no common intellectual framework for co-ordinating results from different areas. Where agreement on goals can be reached, the reduction of technical task uncertainty is sufficient for competing research strategies to be compared in terms of their efficacy in producing the required results but such agreement is often unstable and the outcome of negotiations and conflicts. The combination of problem variety, instability, and multiple orderings of their importance with technical standardization occurs in a number of scientific fields, especially perhaps biological ones.[14]

The third type of intellectual field is perhaps the most commonly discussed and corresponds closest to popular accounts of science. Both strategic and technical task uncertainty are relatively low in such fields so that results are produced in standardised ways and are clearly linked to theoretical goals which are fairly uniform, stable, and ordered into some sort of hierarchy of overall centrality and importance for the field. Both the technical and theoretical environments of practitioners are quite stable and coherent so that research strategies can be formulated with considerable confidence about likely outcomes and how their significance will be assessed by particular groups. Skills are standardized, as are cognitive objects, and the implications of results for theoretical issues are not too difficult to discern and agree on.

Problems are limited in their variability and are ordered in terms of their likely significance so that it seems quite straightforward to select projects to work on. Research strategies almost develop 'naturally' because the overall intellectual structure is stable and coherent. Scientists here have relatively little autonomy from the dominant hierarchy of objectives and priorities, they are highly constrained by standardized work procedures and theoretical structures. While conflicts over the relative importance of sub-fields may arise, especially when resources became scarce, the overall framework is well established and unlikely to change rapidly. Individual and group attempts to alter priorities and negotiate new ones are less likely to succeed in these fields than in sciences where uncertainty is greater. The general susceptibility of collective thought structures and practices to personal or individual organizational pressure and alternatives is low.

The wide acceptance of background knowledge and work procedures in these sorts of fields enable 'anomalies' to be clearly identified as such and their anomalous qualities to be specified, as Kuhn has pointed out.[15] However, this entrenchment of assumptions and procedures in solidaristic communities also renders it unlikely that major 'anomalies' will be admitted if they imply the need to change the dominant goals and beliefs and the social order which supports them. Radical intellectual change – such as is implied by the term 'revolution' – is improbable where work goals and procedures are relatively uniform and stable and the hierarchy of problems and areas is reproduced and controlled by a strong authority structure. Cognitive disorder implies social disorder and this means that the reputational system has broken down, perhaps because of changes in the environment. The more established is the social structure of a scientific field, and the lower the overall uncertainty in the work process, the less likely are 'anomalies' to lead to 'crises' let alone revolutions.

These sorts of variations in levels of task uncertainty across work organizations in general have been associated with differences in the organization and control of work and in authority structures by many writers. These associations can also be expected to occur in reputational organizations like the sciences, and so I shall now discuss the major differences in patterns of the organization and control of research connected with variations in the extent of task uncertainty. Subsequently I will examine relationship between task uncertainty and contextual factors.

VARIATIONS IN THE DEGREE OF TASK UNCERTAINTY AND THE ORGANIZATION OF SCIENTIFIC WORK

The most important organizational corollary of changes in the degree of task uncertainty in scientific fields is the degree, and forms, of co-ordination and control of task outcomes and research strategies. Generally, the lower the degree of task uncertainty, the easier it is to co-ordinate and control work through standardized and formal procedures, reporting systems, and interconnected goals This, in turn, enables work to be planned in advance by people who do not actually do the work and task outcomes to be integrated by similar groups of non-direct production employees. Centralization of control over work processes and purposes is therefore facilitated by a low degree of task uncertainty.

In the sciences these points apply with rather less force because of the institutional commitment to novelty and thus limitation of the degree to which task uncertainty can be reduced. However, fields manifestly do vary in the diversity and stability of their results and strategies, and these variations are associated with differences in the ways that research is organized and controlled. In particular, increasing the degree of task uncertainty means that research strategies and procedures become less standardized throughout the field so that results and topics are less easily compared and co-ordinated with one another. What a particular study has shown, and its implications for other work, are increasingly ambiguous and cannot be effectively communicated across research sites through highly structured and formalized reporting systems. Thus, the organization of research in similar ways to 'mechanistic' bureaucracies[16] becomes less feasible and central control over research strategies, performance standards, and reputations through the formal communication system breaks down. The overall degree of co-ordination and intergration of research is reduced and the control system becomes more personal as task uncertainty grows. Furthermore, the division of the field into separate yet interdependent specialisms which cross local and national boundaries to form subsidiary reputational organizations, in the manner outlined by Hagstrom's notion of 'segmentation',[17] is less probable as co-ordination becomes more uncertain. A decline in the extent and formality of common background knowledge and assumptions renders such differentiation difficult to establish as uncertainty about significance standard grows. These, and other, connections between task uncertainty and the organization of scientific work will now be further explored by discussing technical and strategic task uncertainty under separate headings.

(a) Increasing technical task uncertainty. Increasing technical task uncertainty in scientific fields has a number of implications for the organization and control of research. These concern the importance of local, as distinct from national and international, competence standards and contingencies in the determination of task outcomes; the greater reliance on personal rather than impersonal means of co-ordinating work within and between research sites; the increasingly diffuse and tacit nature of research skills and cognitive objects; the correspondingly discursive and elaborate symbol system for communicating task outcomes and greater variability in how those outcomes are characterized; and the limited size of co-ordination and control units. Generally, growing technical task uncertainty implies greater reliance upon direct and personal control of how research is carried out, considerable local variations in work goals and processes and more informal communication and co-ordination processes.[18]

Work systems where task outcomes are variable, unstable, and difficult to make collective sense of, are only weakly co-ordinated and controlled through common performance standards since outputs are not standardized. As technical task uncertainty increases, it becomes less easy to assess the nature and meaning of particular results collectively so that the allocation of rewards on the basis of contributions to collective goals develops into a rather contingent and idiosyncratic process. Reputations for particular accomplishments vary across employers and national boundaries as the precise nature of results and their implications are difficult to establish and communicate formally. This is especially noticeable in the human sciences where national intellectual reputations are often incomprehensible to cultural élites in other countries. The particular local conditions of knowledge production affect the meaning of task outcomes more when technical task uncertainty grows because standardization of procedures and interpretative norms declines. Thus, knowledge of the particular circumstances in which results were generated becomes important in determining their meaning and significance and so they cannot be compared and assessed in routine ways across research sites or groups. Different cultures and employment units construe the 'same' contributions in varied ways so that they evaluate their merits differently. National and international co-ordination and control of results and reputations is thus limited in fields with high technical task uncertainty as much of nineteenth-and twentieth-century philosophy and literary studies exemplify.

This increasingly contextual and local nature of task outcomes implies that research skills are relatively diffuse and variable across knowledge objects and local cultures. If results are varied, rapidly changing, and open to divergent interpretations, then the skills which produce them and the materials from which they are constructed are not standardized and generalizable throughout the field. Thus local research groups tend to develop distinct ways of conducting and interpreting research on particular sorts of cognitive objects conceived in particular ways. Differences between the Leipzig and Würzburg schools of psychology are an example of this point. As a result, training programmes and skill definitions differ across universities and even more so across national educational systems so that distinct national 'styles' of research become evident in which different sorts of problems and materials are studied with different skills and their outcomes interpreted in different ways.[20]

Not only do skills become more particular to local training organizations as technical task uncertainty increases, but they also often become linked to particular sorts of problems and objects. In contrast to, say, physics where experimental skills are generalizable across a range of problems and topics so that personal mobility is quite considerable,[21] in the human sciences – such as literary studies and historiography – research skills are often tied to certain issues and areas. Here, craft skills are frequently considered in relation to particular cultures, texts, and periods so that they cannot readily be transferred to other ones, let alone to more general and abstract topics as recent debates of the role of 'theory' in English literary studies indicate.[22] Both cognitive objects and skills are delineated in rather diffuse and highly tacit ways so that results are difficult to generalize to other topics or areas. Indeed, the formulation of a particular problem becomes so interconnected with, and dependent upon, extensive knowledge of the phenomena that the nature of research skills are not easy to distinguish from the object of analysis, and competence standards are often particularized around these objects. The high degree of uncertainty about the meaning and implications of task outcomes enables specialists in a particular period or set of texts to dismiss competitors' contributions on the grounds of incompetence or irrelevance as they set performance standards peculiar to that topic. If mutual dependence is not especially high, this can lead to a concern with minutiae and extreme empiricism.

Where skills, cognitive objects, and results are not standardized and are highly mutually implicative, then, co-ordination of task outcomes around collective intellectual goals and the collective award of reputations leading to material rewards cannot be routinized or formalized. Instead, co-ordination and control of research is achieved through personal contacts and knowledge. Just as work-flow variety and instability in general encourage direct, personal supervision of work and co-ordination of tasks,[23] so too scientific fields with a high degree of technical task uncertainty rely heavily on personal networks to make sense of results and assess the merits of contributions.

This, in turn, limits the extent to which a central establishment can directly control and integrate research strategies and results through the formal communication system and ensures considerable local variability of intellectual priorities and significance criteria. Control over how research is conducted and for what purposes it is carried out is quite decentralized when technical task uncertainty is high and dominant coalitions in such fields are typically rather fluid combinations of local leaders connected by largely personal ties and allegiances. Thus teacher – student networks are crucial in co-ordinating research results and in forming distinct political groupings.[24] Additionally, this reliance on personal connections does, of course, restrict the size of sub-units and of the overall reputational system so that rapid expansion of knowledge producers in a field either results in fragmentation or else modifies the degree of task uncertainty so that more standardized ways of working and comparing results of research on more standardized objects are established. The efflorescence of distinct and uncoordinated theory groups in North American sociology after the expansion of the 1960s is an instance of this point.[25]

The importance of personal contacts, and knowledge of the local circumstances in which research is conducted, for the co-ordination and assessment of task outcomes is reflected in the symbol system used to communicate results and persuade colleagues of their correctness and significance. Because objects, skills, and outcomes are not highly standardized and specific, languages in fields with high technical task uncertainty have to be able to convey how research was undertaken, its purpose and implications in greater detail than where tasks and outcomes are more standardized. The high degree of ambiguity of results means that their presentation has to be more elaborate and has to justify the particular interpretation being put forward. Thus typical articles are quite long and often work is communicated through books, whereas a lower degree of task uncertainty enables research to be effectively communicated in a short space through esoteric and standardized symbol systems.

Languages for convincing colleagues are also more personal and variable across scientists where technical task uncertainty is high. Because the meaning and significance of results are ambiguous and uncertain researchers have to persuade others of the correctness of their understanding of the problem and way of dealing with it. The relatively limited degree of standardization of cognitive objects and work processes means that they cannot assume that their views and practices make sense to colleagues. Scientists therefore need to present their work in such a way that the particular approach and interpretation adopted seem convincing. This not only results in reports being quite extensive but also their style has to be tailored to the particular message being communicated. Indeed, how the outcome of a piece of research is expressed constitutes a major part of what it is taken to be in these fields. Just as what and how research is undertaken are more personal and variable, so too is its manner of reporting. Competence in such presentation skills varies, of course, and so affects collective assessments more than in fields where symbol systems and standards governing reporting styles are less tacit and diffuse. Thus, an individual's style of writing is an important component of his or her reputation in the human sciences.

The relatively low degree of standardization and restriction of cognitive objects, skills, and interpretative norms associated with a high degree of technical task uncertainty, and the corresponding importance of personal and local contingencies and practices in determining strategies and outcomes, mean that scientists cannot easily rely upon extra-local colleagues' results in making their own contributions to collective intellectual goals. Nor can they find it straightforward to demonstrate how their particular products to fit in with those of such colleagues to constitute major contributions. Thus, reputations for relatively narrow and specific contributions cannot be realistically expected from those who do not share the particular approach and practices prevalent at the site where they were produced. Highly specialized results, then, usually only produce positive reputations among those who were trained in the same school or have worked in the same group, but not so much among other groups. Consequently, knowledge claims from research groups or schools are wide-ranging and broad in coverage rather than specific and restricted in scope. Each group seeks reputations for their overall contributions to general intellectual goals rather than for how their specific results fit in with the goals and approaches of others.

Because of the diffuseness of evaluation criteria and

variability of interpretative procedures across research sites, then, the dominant unit of intellectual production and evaluation in fields where technical task uncertainty is high is the locally based research group or school. These groups produce rather diffuse and broad contributions with distinct intellectual approaches and procedures and with predominantly local control of critical resources.[26] Their degree of internal cohesion and external exclusiveness varies according to the intensity of competition for control of the reputational system, but they form the major source of identity and co-ordination however limited such connectedness may be. Their strong reliance upon personal loyalties and ties for co-ordinating and making sense of individual contributions limits their size so that a rapid expansion of posts and facilities can result in their dissolution as effective means of integrating task outcomes. Examples of such schools in the human sciences are the 'Chicago school' and the 'structural-functionalist' school in North American sociology and the Würzburg school in German psychology.[27]

(b) Increasing strategic task uncertainty. Just as diversity and variability of task outcomes limit the extent of formalization of co-ordination and control procedures, so too diversity and variability of problems and significance criteria imply a lack of standardized means of co-ordinating and integrating research strategies around common goals. They also imply some variability in the assessment of the worth of contributions and hence in the award of reputations. Thus, a high degree of strategic task uncertainty is associated with considerable organizational fluidity, a lack of hierarchical ordering of problems and goals, limited centralization of control over significance standards, and a considerable degree of theoretical pluralism.

Where intellectual problems vary considerably in their formulation, orientation, and perceived significance, and there are few general criteria for assessing the merits of particular strategies which are widely agreed and stable, scientists are faced with a highly uncertain intellectual environment when deciding what to work on and how to approach problems. While they may share technical procedures and have relatively standardized ways of comparing and co-ordinating task outcomes, the choice of a particular research project or programme reflects a set of intellectual priorities which may not be shared by colleagues and so may not lead to high reputations. The importance of results has to be negotiated and demonstrated rather than being assured by the dominant theoretical structure. Because significance criteria are not standardized and stable, how a given study fits in with, and is relevant for, others' strategies is uncertain and liable to rapid change. Reputations are, then, variable and unstable as competing groups attempt to convince one another of the priority of their problems and approaches, and the superiority of their goals.

This lack of stability and predictability of reputational norms enhances local influences and exigencies in the determination of research strategies. Where the reputational organization has a relatively diverse and rapidly shifting goal structure, individual and group autonomy from central direction and co-ordination is considerable. So researchers and employers are able to pursue distinct strategies and orientations without being penalized for theoretical deviance. As a result, the choice of problems and topics for research becomes quite diverse across research sites and open to a variety of factors. This has become especially noticeable in the bio-medical sciences in recent decades as the former disciplinary structures have declined in importance and research goals are set by what Knorr has termed 'trans-epistemic' arenas.[28] These fields manifest a high degree of fluidity in their priorities and strategies are strongly affected by local exigencies as significance criteria fluctuate and reputational boundaries weaken.

The lack of a clear and generally accepted set of significance criteria where strategic task uncertainty is high means that specialist groups pursuing separate goals disagree over the importance and centrality of problems and there is no commonly agreed hierarchy of topic areas such that reputations in one are judged superior to those in another. Depending on the degree of competition in the field, this may result in researchers forming highly specialized sub-groups around particular concerns which are relatively uncoordinated or unconnected as in US mathematics.[29] In any event, the degree to which results and contributions from members of different groups are integrated into a single intellectual structure is limited and the implications of their work for each other difficult to determine. The reputational organization, then, is relatively decentralized in its political structure with considerable goal diversity and unstable co-ordination processes. As strategic task uncertainty increases, the ability of a central group or school to control and co-ordinate research strategies throughout the field declines since significance standards diverge and become the basis of separate groups forming to produce knowledge claims oriented to distinct problems.

This decline in central control over research goals is linked, of course, to the limited extent of shared background knowledge and assumptions. Growing strategic task uncertainty implies an increase in uncertainty about the general implications of results and task outcomes for collective goals so that research is evaluated differently by different groups. Its theoretical meaning and significance is open to dispute and alternative views so that a common, integrated, and coherent theoretical structure which co-ordinates and interrelates results across groups and problem areas is improbable. Instead, different groups pursuing different intellectual priorities will develop distinct theoretical orientations so that the overall level of theoretical diversity becomes greater as strategic task uncertainty increases. Thus, variations in theoretical position are more legitimate and observable in some biological fields than in mid-twentieth-century physics.

Such growth in theoretical diversity means that sub-unit formation is not just based upon the pursuit of different problems and different rankings of such problems as strategic task uncertainty increases, but also follows alternative theoretical stances, at least where competition for reputational control encourages the development of general implications and theoretical claims. Problems not only differ in perceived importance but also in the way they are formulated and understood so that preferred explanations vary across groups and the ways in which task outcomes are co-ordinated and generalized also differ. As long as technical standardization is sufficient to enable results to be compared and contrasted, this should not lead to sectarian fissions but separate schools may well form around opposed conceptions of the central issues and preferred ways of tackling them so that co-ordination and integration of their research becomes very difficult. The various schools in behaviourist psychology in the USA seemed to exemplify this situation,[30] since they reacted to one another's results and sought to incorporate them in their own work but followed different theories and problem formulations.

In summary, variations in the degree of task uncertainty in scientific fields are associated with differences in the standardization of skills, cognitive objects, significance criteria, and communication structures. They are also related to the degree of centralization of control over the reputational system, the extent to which research is organized and assessed by non-direct production workers, and the importance of personal connections in co-ordinating work and strategies. Finally, they are connected to the sort of sub-units that form in scientific fields, their size and degree of mutual importance. Some of these relationships are similar to those discussed in the previous chapter, especially in connection with skill standardization. Indeed, it seems improbable that a high degree of functional dependence can develop in a scientific fields without task outcomes being relatively standardized and hence technical task uncertainty limited. Thus, the former implies a reduction in the latter. This, and some other connections between these two general dimensions, will be discussed in more detail in the next chapter. Here, however, I shall outline some of the major contextual features which are associated with variations in task uncertainty in a similar way to that adopted in the previous chapter.

CONTEXTUAL FACTORS AFFECTING THE DEGREE OF TASK UNCERTAINTY IN SCIENTIFIC FIELDS

Variations in the degree of task uncertainty in the sciences are constrained by two general forces which act in opposing directions. First, there is the institutional goal of reducing uncertainty so that greater control can be exerted over the environment. Second, there is the professional need to maintain sufficient uncertainty in the process of knowledge production to avoid routinization and external control of the research process. A major justification for the support of much scientific research in modern industrial societies has been its ability to produce reliable knowledge of various environments which assisted in the management and control of them. The extension of public support for many sciences in the nineteenth century was premised upon their capacity to reduce uncertainty in key areas, such as agriculture,[31] and increase human control. Scientists have claimed unique powers of uncertainty reduction as the basis of their legitimacy and demands for support; they therefore have to increase certainty in a demonstrable manner.

However, scientists are also constrained by their professional self-interest to retain control over how such uncertainty reduction takes place and so over knowledge production skills. They additionally seek to augment professional autonomy by restricting the ability of clients to judge and assess the value of task outcomes and so control performance standards. Just as medical practitioners try to control the social meaning of health and illness, so too scientists attempt to monopolize the assessment of knowledge claims and their reliability. Success in reducing uncertainty is decided by professional colleagues in many of the public sciences, rather than by employers or 'clients', and so collegiate control over how work is carried out, and for what purposes, is maintained. Scientists have, then, sought to monopolize control over both the definition, production, and certification of research skills, and the evaluation of task outcomes in terms of their validity and significance. Uncertainty reduction is a key goal of the modern public sciences but one which is interpreted and elaborated largely by scientists. As a result, knowledge claims are evaluated according to their reliability and validity on the one hand, and their usefulness to colleagues on the other. Their value is largely determined by the internal dynamics of the scientific enterprise which is oriented to the production of collectively controlled and ordered innovations. They therefore simultaneously order experiences and provide material for further innovations. Uncertainty is reduced by scientific research, then, but in such a way as to generate more uncertainty for resolution in an unending process of knowledge construction and modification.[32]

This general feature of knowledge production in the modern sciences is manifested in different ways in different fields depending on their environment. The degree to which task outcomes are visible, replicable, and stable, and are organized into coherent theoretical structures, clearly varies between fields and is related to particular aspects of their intellectual and social environment. These can be distinguished in the same way as in the previous chapter into three sets of contextual factors: reputational autonomy, concentration of control over the means of intellectual production and dissemination, and the diversity of audiences.

(a) Reputational autonomy. Given the importance of demonstrating technical control over phenomena for acquiring public legitimacy and resources, high reputational control of competence and significance standards is likely to be associated with considerable reductions in the level of technical task uncertainty. Where such technical control is difficult to manifest, the ability of researchers to control competence and performance standards totally is limited and in general it seems reasonable to suggest that growing technical certainty is a necessary, though not sufficient, condition for growing reputational control over these standards. Thus fields where competence is assessed by employers, the lay public and scientists from other areas, in addition to reputational leaders, are unlikely to develop a high level of technical control with esoteric and standardized techniques since results can always be challenged by those adhering to different standards. The subservience of much work in management studies to standards set by economists, psychologists, and managerial 'practitioners' is an instance of this point.[33]

Low reputational autonomy is also likely to restrict the reduction of strategic task uncertainty since this implies the ability of reputational leaders to order problems in a particular way and ensure the dominance of a given set of significance criteria. Where they share this setting of priorities with other groups, such as funding agency directors and general cultural élites, scientists obviously have a greater degree of latitude from a particular group of reputational leaders in selecting research strategies and so significance criteria are more fluid and varied. On the other hand, however, it does not necessarily follow that increasing the degree of autonomy over the setting of competence and significance standards encourages greater theoretical unification and uniformity of problems. While such autonomy is probably a necessary condition for such developments, it is by no means a sufficient one. Where scientists have substantial control over work processes and goals, they may pursue distinct intellectual priorities and approaches without needing to integrate these into a single intellectual structure, as the case of experimental physiology demonstrates.[34] The internal political systems of fields with considerable reputational autonomy vary in their degree of hierarchy and so too does the degree of strategic task uncertainty.

The extent to which a reputational organization is able to control descriptions of intellectual objects and admissible concepts for understanding them is a further critical factor in the determination of levels of task uncertainty. Where this is low, precise boundaries and sharp distinctions are difficult to develop and maintain, and it is improble that technical control over cognitive objects will be easy to achieve or demonstrate. This is because any attempt at closing and restricting the meaning of phenomena can be opposed by others on grounds of relevance or of omitting the 'real' nature of the object. Standardization of the raw material to be transformed in scientific research requires the limitation of features and characteristics to essential and tractable ones. This limitation can always be challenged in fields where commonsense descriptions and accounts are available and have some legitimacy. by reference to properties and aspects which seem important in those accounts. Psychological restrictions of the concept of individual personality, for instance, have been attacked by those who claim that they fail to deal with the most crucial aspects of individuals. Similar points have been made by critics of social psychological studies of small groups which restricted them to laboratory situations. Where such attacks are regarded as legitimate, they reduce the dominance of technical, restricted conceptions of objects and the ability to generate and establish technical standardization of descriptions and procedures. Closeness to commonsense descriptions makes the formulation and imposition of purely technical ones for scientific purpose difficult. A prerequisite for technical standardization and reduction of technical task uncertainty, then, is some separation of cognitive objects from commonsense formulations.

This point applies even more strongly to the reduction of strategic task uncertainty. Where lay and everyday goals and concepts are important influences on the selection and formulation of scientific objectivies and conceptual approaches it is most improbable that a scientific field will be able to develop a consistent and coherent ordering of intellectual objectives such that contributions can be evaluated in a straightforward and relatively uncontentious manner. The more diverse and diffuse are cognitive objects and purposes, which is probable when commonsense reasoning and language influences scientific thought to a considerable extent, the less likely are scientists to agree on priorities and the relative significance of results. In fields with little autonomy from commonsense reasoning and goals, objectives and concepts are more likely to be imprecise, divergent, and fluid. A high degree of theoretical integration and coherence is, therefore, improbable as is demonstrated by much Anglo-Saxon sociology and political science.

(b) Concentration of control over the means of intellectual production and dissemination. When oligarchic control of reputational organizations is reinforced by concentrated control over access to the means of production some technical standarization and formalization of the reporting system is encouraged so that research can be centrally co-ordinated. Since a high degree of concentration of control over critical resources increases the degree of dependence in a field, scientists will attempt to follow the dominant way of doing research and linguistic style for communicating task outcomes. This does, though, assume the size of the field is so large as to require intellectual co-ordination to be accomplished through the communication system as well as through personal contacts

and knowledge. A high degree of concentration can also occur, however, when the number of knowledge producers is quite small and controlled largely through personal patronage, as in the early nineteenth century French sciences.[35] Here, formalization of the symbol system and a high degree of technical standarization are less necessary since personal ties and connections function as co-ordination mechanisms. Control over research strategies and comptence standards is exercised directly through personal supervision and the allocation of jobs and other resources to favoured disciplines. Thus, results are integrated through patrimonial rather than professional authority structures. Technical task uncertainty can, therefore, remain considerably greater than when the reputational organization is larger and more dependent upon the formal communication system for co-ordinating task outcomes and allocating rewards through collective reputations.

In general terms, where researchers are personally dependent upon a few leaders who directly control resources through administrative positions and patronage, in addition to their influence on reputational goals and standards, rather than upon the reputational system as a whole functioning in a largely anonymous fashion, the degree of technical task uncertainty remains considerable because the dominant unit of knowledge production is the personally controlled research school. Communication to, and persuasion of, specialist colleagues in other schools are secondary to obtaining the approval and support of one's patron and so a high degree of standardization of objects, procedures, and results is not especially important. Indeed, where such schools are based in separate employment organizations which control most of the required resources for knowledge production, technical task uncertainty in the fields as a whole – as distinct from that within each school – is likely to be quite high. So a high level of concentration of control over resources within personally controlled schools combined with relatively decentralized control of resources between schools is associated with considerable levels of uncertainty, ambiguity, and disagreement about the meaning and significance of results. The power of the *Ordinarius* professor in the German university system is an example of such high 'local concentration' of control over key resources which led to distinct schools being formed in the nineteenth century.[36]

Increasing technical control of knowledge objects and standardization of work procedures becomes more attractive and probable when the power of such local barons is reduced and scientists are more oriented to their specialist colleagues. If they have to convince memebers of the field in general, rather than the leaders of their school, of the correctness and merits of their contributions then they are more likely to develop common procedures, skills, and interpretive standards. Indeed, Collins has suggested that where there are a substantial number of researchers seeking reputations they will seek to establish universalistic standards of competence and undertake empirical work as a means of escaping from patron control and increasing their own autonomy.[37] This does, of course, assume that such escape is feasible and does not apply quite so much to European employment organizations as perhaps to some North American ones. Certainly where authority is more equally distributed in universities, and resources are allocated on a national or international level through such mechanisms as peer review panels and referring contributions for publications, it seems probable that technical standardization will be encouraged as scientists become more dependent upon the reputational system than upon their personal connections and knowledge. Thus, the increasing importance of federal funds for research in many of the enlarged human sciences may have resulted in some reduction of technical task uncertainty over the present century in the USA.[38]

A low degree of concentration of control over resources between research groups is also likely to encourage a considerable degree of strategic task uncertainty as scientists claim originality by working on distinct problems with distinct approaches. Where many employment units have control over necessary research facilities and funds, or there is relatively little difficulty in obtaining them from variety of funding agencies, researchers can pursue a variety of goals and priorities and adhere to different significance standards. Once entrenched, such autonomy is likely to be surrendered without resistance, especially if that would imply a devaluation of their specialism and skills. In these circumstance any integrating theory which purports to organize problem areas and conceptual approaches is not going to be accepted very easily, as the fate of general systems theory in the biological and social sciences indicates.[39] General theories which promise to co-ordinate and integrate work in a variety of sub-fields are not popular in sciences where control is relatively widely dispersed because they reduce intellectual autonomy and threaten specialist identities. Problem uniformity and stability are therefore improbable where control of critical resources is broadly diffused and researchers not highly dependent upon a single agency or highly prestigious journal. The debates over Darwin's theory in nineteenth-century natural history reflect this unwillingness to accept new general theories in established fields with specialist skills.[40]

(c) Audience plurality and diversity. Just as a high degree of audience diversity reduces the degree of mutal dependence in a scientific field, so too it reduces the need to develop common, standardized methods of working and communicating task outcomes. Where researchers can legitimately address their results to a number of different groups for reputations they will be encouraged to produce knowledge claims which fit the particular interests and procedures of these separate groups and so become less likely to standardize languages and objects throughout the fields. This is even more probable, of course, if lay audiences are regarded as legitimate addressees and sources of reputations since commonsense descriptions of objects and interpretative practices, which increase the variability and diffuseness of task outcomes, are then legitimated. In general, then, the more diverse and rapidly changing are possible audiences for scientists' work the greater the level of task uncertainty in a scientific field. This is exemplified by many of the human sciences, especially when they appealed to general cultural élites. As audiences have become more narrowly academic and professionalized, so too have scholars tended to become more technical and esoteric in their methods and languages.[41]

Even when the lay public has been effectively excluded from the competent consideration of research results, audience variety can still be a significant factor in determining the level of task uncertainty in scientific fields. This is especially so in respect of strategic task uncertainty. In the same way that a wide distribution of control over resources encourages problem diversity, so too the availability of a variety of possible audiences, which are not very different in terms of the prestige of their reputations, tends to increase the level of strategic task uncertainty. Where scientists are able to publish their results in a number of journals addressed to distinct audiences they will obviously have greater latitude in formulating research strategies than if they had to focus their work on one particular group or if the prestige of reputations in one area was much more than those of other groups. In the bio-medical sciences, for instance, the plurality and diversity of audiences seems much greater than in, say, physics, and so too is the level of strategic task uncertainty. Researchers have considerable scope in deciding what problem to work on, where to publish results, and which audience to impress; furthermore their preferences in these respects are quite fluid and liable to change.[42] A similar situation seems to exist in artificial intelligence where common craft skills are used for a variety of goals and topics which are of interest to a wide variety of audiences and are not ordered into a common theoretical structure.[43]

In conclusion, then, a low level of reputational autonomy, high reliance on personal means of directing and co-ordinating research, and openess to lay audiences are likely to be associated with considerable task uncertainty. The more important general reputations through the field become, and the lower the diversity of audiences, the more technical control and competence will be emphasized and interpreted in terms of common standards. Strategic task uncertainty is also encouraged by a low degree of reputational autonomy, low concentration of control of resources between employment organizations and increasing diversity of audiences. It is likely to decline as the availability of resources is reduced and access becomes concentrated through a single agency or élite group, and as journals and audiences become ordered into a single hierarchy of prestige controlled by a single set of significance criteria.

SUMMARY

The points made in this chapter can be summarized in the following way:

1. The degree of task uncertainty varies considerably between fields and is related to differences in their patterns of work organizations and control. It also varies in relation to changing contextual factors.
2. Task uncertainty has two major aspects: technical and strategic. Technical task uncertainty refers to the visibility, uniformity, and stability of task outcomes. Strategic task uncertainty refers to the uniformity, stability, and integration of research strategies and goals.
3. A high level of technical task uncertainty usually implies a high level of strategic task uncertainty but the reverse is not necessarily so.
4. Increases in the degree of technical task uncertainty limit the size of reputational organizations, lead to greater reliance on personal, direct control of research and its co-ordination, restrict the degree of standardization of skills, raw materials, and symbol systems, and encourage diffuse, discursive intellectual contributions.
5. Increasing the degree of strategic task uncertainty is associated with greater theoretical diversity, reductions in the degree of central control over research goals, and increasing local autonomy in the formulation of strategies and significance standards.
6. Technical task uncertainty is higher when lay audiences and interest groups can influence competence standards and problem formulations, and when personal patronage is the major means of control over work.
7. Strategic task uncertainty is higher when there are a variety of funding agencies and audiences, and no single hierarchy of prestige or significance standards has become entrenched in the resource allocation system.

Notes and references

[1] Among the better known discussions are T. Burns and G. M. Stalker, *The Management of Innovation,* Tavistock, 1961 and J. Woodward, *Industrial Organisation,* Oxford University Press, 1965.

[2] Cf. C. Perrow, *Organizational Analysis*, London: Tavistock, 1970, pp. 80–2.

[3] These are usefully summarized by Perrow, *idem*, 1970, pp. 75–85. See also R. Collins, *Conflict Sociology*, New York: Academic Press, 1975 pp. 321–9; H. Mintzberg, *The Structuring of Organizations*, Englewood Cliffs, New Jersey: Prentice-Hall, 1979.

[4] E. g. E. Harvey, 'Technology and the Structure of Organization', *American Sociological Review*, 33 (1968), 247–58; C. Perrow, 'A Framework for the Comparative Analysis of Organization', *American Sociological Review*, 32 (1967), 194–208.

[5] Perrow, op. cit., 1970, note 2, pp. 78–80.

[6] M. Polyani, *The Tacit Dimension*, London: Routledge & Kegan Paul, 1966; J. R. Ravetz, *Scientific Knowledge and its Social Problems*, Oxford University Press, 1971, pp. 76–108; B. Latour and S. Woolgar, *Laboratory Life*, London: Sage, 1979; K. Knorr, 'Tinkering toward Success: Prelude to a Theory of Scientific Practice', *Theory and Society*, 8, 347–76.

[7] A. Rip, 'The Development of Restrictedness in the Sciences', in N. Elias *et al.* (eds.), *Scientific Establishments and Hierarchies*, Sociology of the Sciences Yearbook 6, Dordrecht, Reidel, 1982.

[8] See, for example, James D. Thompson, *Organizations in Action*, New York: McGraw-Hill, 1967 and H. Aldrich and S. Mindlin, 'Uncertainty and Dependence: Two Perspectives on Environment', in L. Karpik (ed.), *Organization and Environment*, London: Sage 1978.

[9] E. g. Perrow, op. cit., 1970, note 2.

[10] A. Stinchcombe, 'Bureaucratic and Craft Administration of Production', *Administrative Science Quarterly*, 4 (1959), 137–58; Perrow, op. cit., 1970, note 2, 77–8.

[11] As reported in the recent SSRC evaluation of forecasting models. See M. J. Artis, 'Why Do Forecasts Differ?' *Bank of England Paper No. 17* 1982; S. Weir, 'The Model that Crashed', *New Society*, 12 (August 1982), see also, E. A. Leamer, 'Let's Take The Con out of Econometrics', *American Economic Review*, 73 (1983), 31–43.

[12] P. Deane, 'The Scope and Method of Economic Science', *The Economic Journal*, 93 (1983), 1–12 at p. 8.

[13] See H. Katouzian, *Ideology and Method in Economics*, London: Macmillan, 1981, pp. 165–72 and P. Jenkin, *Microeconomics and British Government in the 1970s: The Application of Economic Rationality to Transport, Manpower and Health Policy*, unpublished PhD thesis, Manchester University, 1981, chs. 2 and 3; S. Pollard, *The Wasting of the British Economy*, London: Croom Helm, 1982, ch. 7.

[14] At least as documented by Latour and Woolgar, op. cit., 1979, note 6, and K. Knorr-Cetina, *The Manufacture of Knowledge*, Oxford, Pergamon, 1981.

[15] T. S. Kuhn, *The Structure of Scientific Revolutions*, Chicago University

Press, 2nd ed., 1970; 'The Function of Measurement in Modern Physical Science', in *The Essential Tension*, Chicago University Press, 1977.

[16] As discussed by Burns and Stalker, op. cit., 1961, note 1, pp. 199–225.

[17] W. O. Hagstrom, *The Scientific Community*, New York: Basic Books, 1965, ch. 4.

[18] The personal and fundamental nature of philosophical theories in the early twentieth century United States is linked to the lack of consensus and coherence in philosophy before the 1920s by D. J. Wilson, 'Professionalisation and Organised Discussion in the American Philosophical Association, 1900–1922', *Journal of the History of Philosophy*, 17 (1979), 53–69.

[19] See M. Ash, 'Wilhelm Wundt and Oswald Külpe, on the Institutional Status of Psychology', in W. D. Bringmann and R. D. Tweney (eds.), *Wundt Studies*, Toronto: Hogrefe, 1980. Compare G. Böhme, 'Cognitive Norms, Knowledge Interests and the Constitution of the Scientific Object', in E. Mendelsohn *et al.* (eds.), *The Social Production of Scientific Knowledge*, Sociology of the Sciences Yearbook 1, Dordrecht: Reidel, 1977.

[20] For a discussion of different intellectual styles, see J. Galtung, 'Structure, Culture and Intellectual Style: an essay comparing saxonic, teutonic, gallic and nipponic approaches', *Social Science Information*, 20 (1981), 817–56.

[21] As recorded by Hagstrom, op. cit., 1965, note 17, pp. 160–1; compare B. Harvey, 'The Effect of Social Context on the Process of Scientific Investigation', in K. Knorr *et al.* (eds.), *The Social Process of Scientific Investigation*, Sociology of the Sciences Yearbook 4, Dordrecht: Reidel, 1980.

[22] See, for instance, the symposium on 'Professing Literature' in the *Times Literary Supplement*, 10 December 1982 and extensive subsequent correspondence.

[23] As already discussed in Chapter 2; compare Stinchcombe, op. cit., 1959, note 10.

[24] As discussed by Mullins in his account of North American sociology, see N. Mullins, *Theories and Theory Groups in Contemporary American Sociology*, New York: Harper & Row, 1973. See also N. Wiley, 'The Rise and Fall of Dominating Theories in American Sociology', in W. E. Snizek *et al.* (eds.), *Contemporary Issues in Theory and Research*, London: Aldwych, 1979.

[25] As chronicled by Mullins, *idem*.

[26] The importance of local control of jobs is evident in British social anthropology. See A. Kuper, *Anthropologists and Anthropology*, Harmondsworth: Penguin, 1975.

[27] As documented by Mullins, op. cit., 1973, note 24; Wiley, op. cit., 1979, note 24; E. A. Tiryakian, 'The Significance of Schools in the Development of Sociology', in W. E. Snizek *et al.* (eds.), *Contemporary Issues in Theory and Research*, London: Aldwych Press, 1979 and M. Ash, 'Academic Politics in the History of Science: Experimental Psychology in Germany, 1879–1941', *Central European History*, 14 (1981), 255–86.

[28] K. Knorr–Cetina, 'Scientific Communities or Transepistemic Arenas of Research?' *Social Studies of Science*, 12 (1982), 107–30.

[29] Hagstrom, op. cit., 1965, note 16, pp. 228–35. Compare L. L. Hargens, *Patterns of Scientific Research*, Washington D. C.: American Sociology Association, 1975, ch. 2.

[30] As discussed by Brian Mackenzie, *Behaviourism and the Limits of Scientific Method*, London: Routledge & Kegan Paul, 1977, ch. 1. However, they varied in their experimental procedures and raw materials, i.e. the strain of rat, so that it is not clear that Hull and Tolman were actually producing comparable results. See R. L. Rosnow, *Paradigms in Transition*, Oxford University Press, 1981, pp. 100–2.

[31] For an account of how support for the Royal Institution was linked to agricultural interests and encouraged Davy's work in agricultural chemistry, see M. Berman, *Social Change and Scientific Organisation, 1799–1844*, London: Heinemann, 1978, chs. 1, 2, and 3. On the changing public rhetoric and polemics in support of science in the nineteenth century, see F. M. Turner, 'Public Science in Britain, 1880–1919', *Isis*, 71 (1980), 589–608.

[32] For a brief discussion of how uncertainty and ignorance are constructed in scientific research, see M. Callon, 'Struggles and Negotiations to Define What is Problematic and What is Not', in K. Knorr *et al.* (eds.), *The Social Process of Scientific Investigation*, Sociology of the Sciences Yearbook 4, Dordrecht: Reidel, 1980.

[33] Compare R. D. Whitley, 'The Development of Management Studies as a Fragmented Adhocracy', *Social Science Information* 23, 1984.

[34] In the nineteenth century at any rate. See G. Allen, *Life Science in the Twentieth Century*, Cambridge University Press, 1978, pp. 82–94; G. Geison, *Michael Foster and the Cambridge School of Physiology*, Princeton University Press, 1978, pp. 331–51.

[35] On the powers of the *grands patrons* in the early ninetennth century and their use of patronage to control research, see R. Fox, 'Scientific Exterprise and the Patronage of Research in France, 1800–70', *Minerva*, 11 (1973), 442–73; 'The Rise and Fall of Laplacian Physics', *Historical Studies in the Physical Sciences*, 4 (1974), 89–136; 'Science, the University and the State', in G. Geison (ed.), *Professions and the State in France*, University of Pennsylvania Press, 1984.

[36] Such as those in psychology, see M. Ash, op. cit., 1981, note 27.

[37] R. Collins, op. cit., 1975, note 3, pp. 510–11.

[38] For a brief discussion of such changes in biology, see M. Heirich, 'Why We Avoid the Key Questions: How Shifts in Funding of Scientific Inquiries Affect Decision-making about Science', in S. Stich and D. Jackson (eds.), *The recombinant DNA Debate*, University of Michigan Press, 1977. On the expansion of graduate education and the mathematization of economics after 1950, see H. G. Johnson, 'National Styles in Economic Research', *Daedalus*, 102 (1973), 65–74.

[39] For an illustration of how empirical biologists ignored mathematical model builders during a purportedly inter-disciplinary project, see J. Bärmark and Göran Wallen, 'The Development of an Interdisciplinary project', in K. Knorr *et al.* (eds.), *The Social Process of Scientific Investigation*, Sociology of the Sciences Yearbook 4, Dordrecht: Reidel, 1980. On the fate of the Theoretical Biology Club in the 1930s, see D. Haraway, *Crystals, Fabrics and Fields*, Yale University Press, 1976, pp. 124–35.

[40] For the case of ornithology see P. L. Farber, *The Emergence of Ornithology as a Scientific Discipline, 1760–1850*, Dordrecht: Reidel, 1982, pp. 146–7.

[41] A phenomenon much deplored by some members of English literature departments. See the debate on 'professionalism' in the *T.L.S.*, op. cit., 1982, note 22 and later letters.

[42] As reported in the studies of laboratories by Latour and Woolgar, op. cit., 1979, note 6 and Knorr-Cetina, op. cit., 1981, note 14.

[43] See J. Fleck, 'Development and Establishment in Artificial Intelligence', in N. Elias *et al.* (eds.), *Scientific Establishments and Hierarchies*, Sociology of the Sciences Yearbook 6, Dordrecht: Reidel, 1982.

5

THE ORGANIZATIONAL STRUCTURES
OF SCIENTIFIC FIELDS

In the previous two chapters I have discussed two major dimensions of organizational structure which are related to major differences in how scientific work is organized and controlled in different fields, and to variations in the context of different sciences. Essentially, these dimensions summarize some of the critical features of heterogenous scientific fields considered as distinct reputational organizations, and enable us to explore relationships between these features and the environments of reputational organizations. They thus provide a connection between certain general aspects of the social contexts of scientific fields, and their dominant patterns of knowledge production and validation.

Having outlined these dimensions separately, I shall now discuss their interconnections and explore particular combinations of them as different types of sciences. In this and the next chapter, scientific fields will be characterized in terms of certain interrelationships between these two dimensions and their patterns of development linked to changes in their enviroments. The present chapter will focus on how the degree of mutual dependence and of task uncertainty are related and are associated with the internal structures of different types of sciences, while the next chapter will deal with the contexts of scientific fields and how they change. Initially, I shall discuss how the general dimensions of organizational structure combine to characterize seven major types of scientific field. Subsequently, the variety of organizational patterns found in each of these seven types will be outlined along similar lines to those mentioned in the previous chapters together with some additional features arising from the combination of dependence and task uncertainty. As before, I will illustrate the points being made with examples drawn from the secondary literature where appropriate. These are intended to illuminate and clarify the analytical framework being presented rather than being being systematic accounts of the particular sciences.

Given the division of these two dimensions into two distinct aspects, and their further separation into high and low positions, it would be possible to generate sixteen different types of scientific fields as illustrated in Table 5.1. However, some of these combinations are improbable and others are only likely to develop and some established as distinct organizations in rather special circumstances. An example of the latter is the combination of high technical task uncertainty, low strategic task uncertainty and low mutual dependence.

As discussed in the previous chapter, the existence of reputational organizations combining uniform, stable, and coherently ordered problems and research strategies with variable, unstable, and ambiguous task outcomes is improbable. If the degree of mutual dependence between scientists is, in addition, limited so that they are able to address a variety of audiences for reputations and, hence, pursue different purposes, it is difficult to see how problem selection and formulation remain standarized and restricted. The fissiparous pressures inherent in local variability of research results and a plurality of audiences will be so great as to increase the level of strategic task uncertainty and thus change the type of the field. In the case of Anglo-Saxon economics, which seems to be the only example of this combination of the two types of uncertainty, the level of mutual dependence is rather high because of the uniformity of undergraduate training, the hierarchical communication system, and the high value attached to analytical and theoretical elaboration at the expense of empirical 'applications' of the dominant approach. Low strategic task uncertainty is maintained, here, by strict control over skills and the partitioning of empirical work from the theoretical orthodoxy. Reputations are highest in the analytical core of the field which remains largely immune to empirical uncertainties.[1] As Phyllis Deane suggests: 'the

TABLE 5.1

Possible Combinations of the Degree of Mutual Dependence and the Degree of Task Uncertainty in the Sciences

			Degree of functional dependence			
			Low		High	
			Degree of strategic dependence		Degree of strategic dependence	
			Low	High	Low	High
Degree of technical task uncertainty	Low	Degree of strategic task uncertainty — Low	1	2	3	4
		High	5	6	7	8
	High	Degree of strategic task uncertainty — Low	9	10	11	12
		High	13	14	15	16

leading theorists left little doubt that progress in economic science (as opposed to political economy) could take place only at the theoretical core.'[2] Thus strategic dependence is high and yet the lack of technical control over empirical knowledge objects limits the extent of functional interdependence, except within the central core. In the case of economics, then, a high level of technical task uncertainty is combined with a low level of strategic task uncertainty, considerable strategic dependence, and limited functional dependence between economists in the 'applied' periphery. To quote Deane again: 'questions of empirical validity. . . simply did not arise at the theoretical level. They could be relevant only within the more flexible periphery of applied economics.'[3] Aside from this particular combination, fields combining low strategic task uncertainty with high technical task uncertainty seem unlikely to be stable as reputational organizations.

More generally, a high degree of functional dependence between researchers requires some technical control of phenomena and comparability of task outcomes so that fields with a high level of technical task uncertainty are unlikely to develop much functional dependence. If scientists have to use one another's results to make their own competent contributions to collective intellectual goals, and can only obtain positive reputations by getting colleagues to use their task outcomes, then these need to be comparable and interpreted in similar ways across research sites and schools. This implies some standardization of skills and raw materials so that the level of technical task uncertainty cannot be very high if the degree of functional dependence is considerable. Thus, scientific fields exhibiting the combination of qualities represented by numbers 11, 12, 15, and 16 in Table 5.1 are unlikely to be stable or occur very often. If the degree of dependence in general does increase in a field with limited technical control, scientists are likely to try to standardize skills and knowledge objects, and thus reduce the level of technical task uncertainty.

Equally, the level of technical task uncertainty seems unlikely to be much reduced without some corresponding increase in the level of functional dependence between members of the field. Increasing the degree of standardization of research skills, cognitive objects, and task outcomes reduces the extent of individual and group autonomy over how research is done and encourages co-operation and connections between specialist researchers. It heightens collective reputational control over work procedures and the definition of the particular skills needed to produce competent contributions in particular fields. Relatively low levels of technical task uncertainty imply, then, a considerable degree of co-ordination of work processes across employment units and integration of results. It seems improbable that technical control would increase substantially without national and international reputations growing in importance relative to local ones and so the level of mutual dependence also increasing. Thus scientific fields which reduce technical task uncertainty are also likely to manifest considerable levels of mutal dependence among their members and so the upper left quadrant of Table 5.1, i.e. numbers 1, 2, 5, and 6, is not likely to contain many sciences, except perhaps for those in transition between other, more stable stages.

From the sixteen possible types of scientific fields generated by combining varying degrees of mutual dependence and task uncertainty, then, only seven seem likely to be stable and distinct reputational systems of knowledge production and control. These combine high levels of technical task uncertainty with low levels of functional dependence (numbers 10, 13, and 14) and low levels of technical task uncertainty with high levels of functional dependence (3, 4, 7, and 8). They exhibit varying levels of strategic task uncertainty and of strategic dependence. Each combination of different degrees of these four variables seems to be represented by at least one field and to differ in some important respects from the others. While not claiming that these seven types exhaust all the sciences since they became dominated by employees, or that all fields fit neatly and unambiguously into a single category, this framework does highlight some of the major differences between the sciences and suggest some ways of accounting for them. Rather than simply reproducing current views about, say, the natural and social sciences,[4] the approach presented here offers a more systematic way of distinguishing between such fields as economics and psychology in different periods, and between the sorts of knowledge produced by different national systems of research organization and support, such as the various European traditions in historiography and philosophy. It thus differentiates between scientific fields, between historical situations and between national structures.

TABLE 5.2
Types of Scientific Fields

Degree of task uncertainty	Degree of functional dependence: Low	
	Degree of strategic dependence	
	Low	High
High technical and high strategic task uncertainty	(a) Fragmented adhocracy producing diffuse, discursive knowledge of commonsense objects e.g. Management studies, British sociology, political studies, literary studies, post-1960 US ecology	(b) Polycentric oligarchy producing diffuse, locally co-ordinated knowledge e.g. German psychology before 1933, British social anthropology, German philosophy, Continental European ecology
High technical and low strategic task uncertainty	(c) Unstable	(d) Partitioned Bureaucracy producing both analytical, specific knowledge and ambiguous, empirical knowledge. e.g. Anglo-Saxon economics

	Degree of functional dependence: High	
	Degree of strategic dependence	
	Low	High
Low technical and high strategic task uncertainty	(e) Professional adhocracy producing empirical, specific knowledge e.g. bio-medical science, artificial intelligence, engineering, pre-Darwinian nineteenth century ornithology	(f) Polycentric profession producing specific, theoretically co-ordinated knowledge e.g. Experimental physiology continental mathematics
Low technical and low strategic task uncertainty	(g) Technologically integrated bureaucracy producing empirical, specific knowledge e.g. Twentieth-century chemistry	(h) Conceptually integrated bureaucracy producing specific, theoretically oriented knowledge e.g. Post-1945 physics

In Table 5.2 I suggest some titles and examples of the seven major types of scientific field identified by combining variations in the degree of mutual dependence and of task uncertainty. These titles are intended to summarize some of the important features of each type without implying any particular hierarchy or preference ordering. The term 'adhocracy', for example, is borrowed from Mintzberg's discussion of different organizational types[5] and refers to the political system of reputational organizations rather than to any epistemological account of scientific decision-making. Similarly, 'polycentric' is meant to imply a plurality of power centres and approaches without necessarily suggesting that this sort of political structure is better or worse than any other. I will now outline the main features of these seven types in very general terms before analysing their internal organizations in more detail.

(a) Fragmented adhocracies. The combination of high task uncertainty and low degrees of mutal dependence characterizes what I term 'fragmented adhocracies' because research is rather personal, idiosyncratic, and only weakly co-ordinated across research sites. Individual dependence upon any single reputational organization is limited and so scientists do not have to produce specific contributions which fit in to those of others in a clear and relatively unambiguous manner. Rather, they tend to make relatively diffuse contributions to broad and fluid goals which are highly contingent upon local exigencies and environmental pressures. Typically, these fields are open to the general 'educated public' and have some difficulties in excluding 'amateurs' from competent contributions and from affecting competence standards. The political system is therefore pluralistic and fluid with dominant coalitions being formed by temporary and unstable controllers of resources and charismatic reputational leaders. Reputations are also fairly fluid as standards alter and are open to a variety of interpretations. Given the overall openness of the reputational system, intellectual problems, objects, and procedures are exoteric rather than esoteric and commonsense languages dominate the communication system. Specialization, thus, tends to take place around everyday

empirical objects such as 'education' 'marketing', or 'eighteenth-century English literature'.

(b) Polycentric oligarchies. If the degree of strategic dependence is greatly increased – for instance by centralizing jobs or funds – scientists become much more oriented to the views and intellectual ideas of a small group of intellectual leaders who control scarce resources. Since technical control is still limited, and so results are relatively idiosyncratic to local conditions of their production and interpretation, this control is mostly exercised locally and through personal knowledge. These fields are therefore termed 'polycentric oligarchies' as research is organized into competing schools based on leadership entrenched in employment organizations and control over journals.[6] Knowledge here tends to be more theoretically oriented and co-ordinated than in the previous case, as scientists have to demonstrate the importance of their contribution to the school's overall programme rather than simply claiming reputations on the basis of competence in empirical structures. It is, however, still diffuse because of the highly tacit nature of research skills and lack of standardized interpretative procedures.

(c) Partitioned bureaucracies. The combination of high technical task uncertainty, low strategic uncertainty, and high strategic dependence characterizes 'partitioned bureaucracies' which are highly rule governed and hierarchically organized fields. Here standardiztion of training programmes and skills in the central core enable the reputational élite to control research strategies and problem selection, but the lack of technical control over empirical phenomena – which is the basis of legitimacy claims in the wider social structure – threatens this theoretical coherence and closure. Theoretical elaboration becomes more prestigious than empirical exploration and 'application' of the dominant analytical skills and concepts to empirical objects is partitioned off into sub-fields in a way which does not threaten the prevailing framework and standardized skills. Knowledge in the central core is highly specific and analytical but becomes more ambiguous and empirically oriented in the peripheral 'applied' areas.

(d) Professional adhocracies. Where technical task uncertainty is

more reduced, but strategic task uncertainty remains high, the standard skills and technical procedures enable a more typical 'profession' to develop in which reputational organizations control the production and certification of research competence but differ in the extent to which they control work goals and priorities. In 'professional adhocracies' there are a variety of influences on research goals and no single group dominates significance criteria for very long. The bio-medical sciences and artificial intelligence,[7] for example, have a variety of funding sources and employment organizations where research is conducted, and there is no single reputational group to whom all members of the field are oriented and take into account when developing their research strategies. Instead the 'dominant coalition' of these fields is a temporary and shifting alliance of funding agency officials, established scientists, directors of employment organizations, and representatives of powerful non-scientific groups such as those mentioned by Knorr in her account of 'trans-scientific fields'.[8] Knowledge is highly specific and empirically focussed in such fields with a variety of problem formulations and conceptual approaches linked to particular skills. Generality of both problems and materials is unlikely to be very high and a high degree of theoretical integration improbable.

(e) Polycentric professions. Increasing control of research priorities by a single reputational group—such as the set of research institute directors in the German university system—reduces this variety of influences and the autonomy of the individual researcher to pursue diverse strategies with a multiplicity of extra-local agencies. The resulting high degree of mutual dependence leads to 'polycentric professions' being formed in which relatively standardized skills and procedures are organized around different research programmes and 'schools'[9] typically centred on a small number of employment organizations and leaders. However, because skills are relatively standardized, results are comparable across these schools and contrasting programmes are not so unrelated as in polycentric oligarchies. Collective judgements about the significance of programmes and results are easier to generate and the common background of work procedures ensures that

disputes are not so intense, and unresolvable, as in those fields. Knowledge is here more theoretically oriented than in professional adhocracies as each school justifies its problems and approach in terms of general goals and scientists are more likely to invoke general meta-scientific arguments when competing over intellectual priorities.[10]

(f) Technologically integrated bureaucracies. Where task uncertainty is relatively low on both dimensions, research can be much more rule governed and planned. Standardization of problems and skills encourages a more bureaucratized form of work organization and control as the theoretical framework becomes more stable and precise and embedded in the research technology. Fields which have few problems about the availability of this technology and other resources do not need to compete over the centrality of sub-groups to the collective objectives of their discipline and so scientists are not very concerned about their general contribution to the field as a whole but rather focus on particular sub-problems and sub-goals. Such fields can be termed 'technologically integrated bureaucracies' because results are co-ordinated largely through the research technology which ensures that theory, methods, and phenomena studies are all integrated. Collins considers these sorts of sciences to be similar to Woodward's process industry firms in which the technical system dealt with most of the co-ordination problems resulting in fewer conflicts between sub-groups.[11] Knowledge is highly specific and empirical in these fields, focusing on a relatively large number of properties of particular phenomena rather than highly general and abstract features of 'fundamental objects.'[12]

(g) Conceptually integrated bureaucracies. Where, in contrast, facilities and other resources are relatively scarce and limited in availability, technological co-ordination of results and problems is not sufficient to deal with the allocation problem as sub-groups compete for access to crucial apparatus and funds. Competing claims to general significance within the overall theoretical framework require adjudication by some central authority, which in analogous mass production firms is usually the top management. In these fields co-ordination

problems are resolved by theoretical elaboration and integration of sub-groups' goals into a unified cognitive order which distinguishes quite systematically between the centrality of their concerns to the disciplinary objectives. Although results are relatively predictable in these fields, and the theoretical implications of task outcomes are fairly easy to discern, the concentration of access to, and scarcity of, resources means that the exact theoretical status and significance of a 'department's' output for the enterprise as a whole needs to be determined for efficient resource allocation to continue and so extensive theoretical work is required. Thus I characterize these fields as 'conceptually integrated bureaucracies'. Knowledge here is more abstract and focused on analytical goals than in the previous case and theoretical co-ordination of results is highly valued.[13]

The examples suggested in Table 5.2 raise some important points about the relevant unit of analysis when comparing scientific fields. Depending on how this is decided, examples can differ in their location in this table since some specialisms may be more co-ordinated and standardized than their 'parent' disciplines. In management studies, for example, operations research could be so considered and if such specialisms were taken as the basic unit or analysis then this one would be in category (e), rather than in (a).[14] Similarly, social mobility research in Anglo-Saxon sociology can be regarded as more highly structured than the field as a whole. Given the disjunction between academic disciplines as units of skill production and certification, and scientific fields as reputational organizational controlling the production and validation of intellectual innovations in many of the modern sciences, it is clearly insufficient simply to take university employment unit labels as constituting the elementary analytical unit for our purposes. In any event, this procedure would be unworkable in some fields, such as the post-1950 biological sciences, where labels vary considerably between organizations.[15]

Taking the immediate specialism to be the basic unit of intellectual and social organization also creates substantial problems, however. First, it is not always very clear how these are to be identified across the sciences as many authors have pointed out.[16] Second, the present high degree of intellectual specialization is historically specific and thus temporal comparisons would be difficult to carry out. Third, they are often highly fluid and changeable so that comparisons of their structure and operations as relatively stable work organizations is fraught with obstacles. Finally, because the degree of specialization and intensification of the division of labour varies across the sciences the sort of specialist communities discussed by Kuhn and some sociologists[17] are either relatively unimportant or simply non-existent in some areas.

As far as the employee-dominated sciences are concerned, it seems sensible to focus on the major unit of social organization that controls access to material resources and rewards through collegiate reputations as allocated by the public communication system. This may be equivalent to the dominant unit of skill production, as in academically dominated fields such as many of the human sciences, but, equally, it may not, as in many of the bio-medical sciences and nineteenth-century natural history. The emphasis here is on the organizational unit that links reputations with access to resources and exercises the major influence upon competence and significance standards. Typically, this is the dominant unit of employment and of control over research facilities such as technicians, apparatus, and materials. Although scientists may have to obtain positive reputations from specialist colleagues in, say, operations research or the sociology of education, they also have to convince the controllers of jobs and facilities that their contributions have broader merits for the field as a whole which dominates the allocation of resources. It is usually this broader social unit which provides the institutional basis for intellectual identities and allegiances and which underpins the selection of shorter-term orientations and strategies. The basic unit of analysis here, then, is the major organizational entity controlling access to jobs, facilities, technicians, and other resources through public reputations.

THE INTERNAL STRUCTURES OF SEVEN MAJOR TYPES OF SCIENTIFIC FIELDS

The seven major types of sciences just discussed organize and control knowledge production and evaluation in different

ways which result in different patterns of intellectual organization. The location of a particular field in one type of
science, then, implies a certain way of structuring research
and a certain characterization of its knowledge. If a given field
changes from one type to another, it should also alter its
internal structure and its dominant patterns of intellectual
organization. For example, the growth of state funding of
bio-medical research in the USA, especially for cancer and
heart diseases, encouraged the decline of traditional disciplinary élites and boundaries in many biological fields, such as
physiology and zoology, and so reduced the degree of strategic
dependence among researchers. Much biological research is
today organized as professional adhocracies rather than
polycentric professions as a result and longer-term strategies
seem relatively weakly co-ordinated as witnessed by recent
studies of bio-medical research laboratories.[18] North American sociology is another field where rapid expansion encouraged a decline in strategic dependence in the 1960s and 1970s
thus changing it from a polycentric oligarchy dominated by
various forms of structural functionalism to a fragmented
adhocracy exhibiting much greater intellectual pluralism.[19]
Such reductions in the degree of strategic dependence between
scientists in scientific fields have, then, been associated with
changes in their internal structures and patterns of intellectual
organization. In the rest of this chapter I will outline the
major features of these internal structures and their varying
occurrence in the seven major types of science.

Particular characteristics of the organization and control of
scientific research which varied with differences in the degree
of task uncertainty and mutual dependence were discussed in
Chapters 3 and 4. These included the following dimensions:
the degree to which tasks and skills were divided and
separated, the scope and generality of problems tackled by
individuals and the specificity of task outcomes, the degree of
standardization of skills, procedures, cognitive objects, and
materials, the intensity and scope of conflicts between
scientists, the importance of general theoretical integration
and co-ordination of research, the diversity of goals and
conceptual approaches, the degree of local and individual
autonomy in work processes and goals, the impersonality and

formality of co-ordination and control procedures and symbol systems, the degree of segmentation of fields, and the extent to which they were divided into separate schools of thought and practice. These and other features of the sciences can, I suggest, be summarized under two general headings: first, the configuration of tasks and problems and, second, the means by which, and degree to which, research is co-ordinated and controlled across research sites and groups.

The configuration of tasks and problems in scientific fields refers to the internal arrangement of work processes and goals in different yet related areas. It incorporates the following four dimensions: (a) the degree of specialization and standardization of tasks and materials, (b) the extent to which fields are segmented into distinct specialisms which study different problems with similar procedures and assumptions, (c) the extent to which fields are differentiated into distinct schools of thought and practice which pursue divergent intellectual goals with contrasting approaches, and (d) the degree to which sub-units are ordered and integrated into a hierarchical structure which represents and reproduces a particular set of significance standards.

Increasing specialization and standardization are linked to the production of highly specific task outcomes and reductions in the scope of problems tackled by individual researchers so that intellectual foci and concerns are narrowed and an extensive division of labour develops. Increasing segmentation of scientific fields implies their division into sub-fields which pursue separate goals and problems but share general intellectual orientations and perspectives. Thus these specialisms do not conflict over metaphysical presumptions and ways of doing research but may compete over the significance and importance of their concerns. In contrast, differentiation by research schools implies greater divergencies of views and goals as opposing groups seek to dominate the reputational system and their strategies are, to some extent, mutually exclusive rather than being complementary. Typically, such schools are based on a small number of employment units and personal networks while segments cross employer boundaries and rely more on impersonal means of co-ordinating task outcomes. This contrast can be illustrated by comparing the division of problems between the Jodrell Bank and Cambridge groups in radio astronomy as an example of segmentation, with the competition between the Cambridge school and the Berlin school in experimental physiology in the nineteenth century over problems, priorities, and approaches.[20] Where both segmentation and school-based differentiation are low, there are few stable and strongly bounded sub-units to co-ordinate research around particular problems and so strategies are fluid and fairly broad. Specialisms may form around particular objects, techniques, or problems, but they tend to be unstable and heterogeneous in structure, such as sub-units in Anglo-Saxon sociology and management studies. Finally, just as some functional departments in business firms are more powerful than others, so too are some problem areas regarded as more crucial to reputational goals than others in scientific fields and the degree to which such variations in centrality are ordered into a single dominant hierarchy of significance criteria also varies. Generally, the more one set of such criteria dominates and acts as the means of setting intellectual priorities and awarding reputations throughout the entire field, the more theoretical uniformity and cohesion can be expected and divergencies restricted. This need not mean that results from all areas are interrelated into a logically integrated structure but that reputations are allocated on the basis of a single set of significance criteria which orders sub-fields and their results.

Processes of co-ordination and control of research refer to the dominant means of connecting and integrating scientific work around common problems and concerns and the extent to which such co-ordination constrains and lessens intellectual conflicts. Four distinct aspects can be distinguished. First, there is the degree to which research is organized by impersonal and formal procedures and reporting systems, as distinct from direct, personal supervision and contacts. Second, there is the degree to which research is co-ordinated around certain theoretical purposes and criteria as distinct from being primarily concerned with the exploration of particular phenomena and properties. Problems and strategies can be selected for largely local and specific reasons or for more general and theoretical purposes and so the degree of theoretical co-ordination of topics varies between fields. Third, there are considerable variations in both the scope and, fourth, the degree or intensity of conflicts between individuals and groups of scientists and these are obviously connected to the extent of co-ordination and integration of task outcomes and research strategies. Some disputes, such as those between ethnomethodologists and functionalists in North American sociology, range over a large number of issues and presuppositions while others, such as the controversy over gravitational radiation waves in physics,[21] focus on only a few points of divergence. Similarly, some conflicts are quite bitter and sharp, such as that between the biometricians and Mendelians,[22] while others are more relaxed and diffuse, such as those between the various approaches to the study of cancer.

These characteristics of the internal organization of scientific fields are listed in Table 5.3, together with the degree to which each of the seven major types of science manifests them. These variations are drawn from the arguments presented in the previous two chapters and described the more substantial differences between scientific fields. They are also connected to differences in dominant intellectual styles and patterns of intellectual organization. These sets of differences will now be discussed in greater detail for each of the seven types of science.

(a) Fragmented adhocracies. The dominant feature of these sciences is their intellectual variety and fluidity. As can be seen from Table 5.3, they do not exhibit a stable configuration of specialized tasks or of problem areas, nor do they have strong co-ordinating mechanisms which systematically interrelate results and strategies. Since task uncertainty is high, and scientists' dependence upon a particular group of specialist colleagues is low, the degree of specialization of tasks and procedures is limited, especially across employment units. Similarly, the formation of stable specialist groups which co-ordinate work from different research sites around common objectives is restricted by interpretative variations and the availability of a variety of audiences for legitimate reputations. Where co-ordination of results does occur, it is

TABLE 5.3
Characteristics of the Internal Structure of Seven Major Types of Scientific Field

Types of scientific field	Characteristics of internal structure							
	Configuration of tasks and problem areas				Co-ordination and control processes			
	Specialization and standardization of tasks and materials	Degree of segmentation	Degree of differentiation into schools	Hierarchization of sub-units	Impersonality and formality of control procedures	Degree of theoretical co-ordination	Scope of conflict	Intensity of conflict
Fragmented adhocracy	Low	Low	Low	Low	Low	Low	High	Low
Polycentric oligarchy	Low	Low	High	Low	Low	High	High	High
Partitioned bureaucracy	High in core Medium in periphery	Medium	Low	High	High in core Medium in periphery	High	Low	Medium
Professional adhocracy	High	Medium	Low	Low	High	Low	Medium	Low
Polycentric profession	High	Medium	High	Low	High	High	Medium	High
Technologically integrated bureaucracy	High	High	Low	Low	High	Medium	Low	Low
Conceptually integrated bureaucracy	High	High	Low	High	High	High	Low	Medium

highly personal and linked to local control of resources as in the development of ethnomethodology in California,[23] yet the general availability of resources and openness of the communication system mean that such co-ordination can be avoided. Thus idiosyncratic and individual research strategies and topics are encouraged and the formation of distinct, integrated schools of research is limited. In this situation, the most important influence upon configurations of tasks and problem areas is the employment system and its links to communication media. How jobs and local research resources are allocated and controlled become crucial to knowledge production and evaluation when there is little standardization of work procedures, competence criteria, and significance criteria across employment units. National variations in employment structures play a major role in the organization of intellectual work and the development of contrasting intellectual styles here,[24] and require a brief discussion.

Where employment organizations are relatively democratically structured, as in élite USA universities and some British universities,[25] sub-groups linked to a variety of intellectual concerns and approaches will find it easier to influence appointments and other resource allocation decisions than in countries where a more autocratic and hierarchical system prevails as in the traditional German structure. Thus, Anglo-Saxon departments tend to recruit specialists in particular areas, such as the sociology of education or eighteenth-century English literature, rather than seek the person with the highest reputation in the overall field. Co-ordination of work across these specialist areas is limited since teaching is often differentiated along similar lines and research can be oriented to fellow specialists rather than to general theoretical goals which compete for intellectual domination of the whole field. Local autocracy, on the other hand, when combined with high strategic dependence, tends to promote greater internal cohesion and co-ordination of results and strategies around particular goals and approaches which compete for reputational leadership. The interrelationships between particular task outcomes and specialized topics become of greater significance here as local orthodoxy is enforced. Theoretical connections and implications are emphasized as rival schools conflict over goals and conceptual approaches. Thus German psychology tended to be more theoretically co-ordinated than much Anglo-Saxon psychology in the inter-war years.[26]

In employment systems where permanent appointments are available at a fairly young age and individuals are able to pursue their own interests and approaches, intellectual diversity is to be expected if the resources required for research are widely available and the communication system loosely organized. However, this does not mean that practitioners are all members of stable and distinct sub-units which control major resources. Rather, they seek reputations in distinct areas whose boundaries are diffusely specified and subject to varied interpretations. Skills are not closely tied to these specializations and so scientists can and do move between them over an intellectual career, just as they may choose to make more general contributions to disciplinary goals through syntheses and theoretical frameworks. These latter atempts at obtaining a general reputation are unlikely to be successful in the sense of dominating the research strategies of the entire discipline as long as the separate sub-groups control sufficient resources to reproduce themselves as reputational systems. Often, they become simply another sub-group which is regarded as a legitimate area for members of the field to work in and acquire reputations. This seems to be what has happened to the ethnomethodologists in North American sociology. Thus sub-groups form around both the objects of concern, conceived largely in commonsense terms, and around distinct methodological approaches in fragmented adhocracies.

It follows from the lack of a standardized communication system and high technical task uncertainty that ordering results from different sub-groups into a systematic hierarchy of significance for intellectual goals is difficult and improbable in fragmented adhocracies. Variations in strategies and interpretations across work-places cannot be co-ordinated and integrated through standardized skills and procedures since these are neither precise nor formal enough to overcome communication problems. Consequently, sub-groups pursuing separate concerns are often linked to particular employment organizations and/or co-ordinated through personal contacts and knowledge without being mutually ordered or even strongly interconnected except perhaps through some general and diffuse set of methodological commitments. As long as there is no need to demonstrate the general significance of a topic area to the field as a whole in order to obtain reputations which lead to jobs and other resources, scientists have little incentive to co-ordinate their problems and agree on some basis for comparing the overall significance of their interests. Indeed, personal interests of autonomy and independence probably· encourage them to fight against any such integration and ordering. Schools of research which threaten this autonomy by proposing an intergrating framework, which would differentiate between the importance of sub-units on the basis of their relative contribution to the school's goals, are therefore likely to be strongly resisted.

The variety of audiences and research strategies in fragemented adhocracies is matched by the variability of sub-unit structure. Some problem areas may be highly co-ordinated around particular skills or techniques – such as laboratory experiments in physiological psychology – while others are much more loosely structured and reliant upon personal means of co-ordination – such as therapeutic psychology. Thus we can extend the two dimensions of task uncertainty and mutual dependence to organizational sub-units and distinguish between those which have attempted to reduce uncertainty by narrowing foci of concern and legitimate research practices – as in much of social mobility research – from those which insist on the highly tacit, personal, and diffuse nature of research skills. Equally, there are some sub-fields where practitioners are quite mutually dependent – such as the psychology of memory and perception – and others where the degree of mutual orientation is much less – such as organizational psychology. If these variations are very large and if the separate groups are able to control resources through their sub-field reputations rather than being mediated by disciplinary reputations, it is doubtful if the field exists in any real sense as the basic unit of intellectual production and integration. Rather, it may function more as a convenient location of jobs and facilities for different reputational groups to claim according to their own ideals and criteria. To some extent, management studies seems to function this way with the sub-fields varying in structure from the highly coherent and restricted groups of finance theorists, who are in many ways a subset of micro-economics, to the very loose and diffuse set of activities and knowledge labelled 'organizational behaviour'. However, their common location in business schools or departments of management 'sciences' limits such fragmentation.[27]

Where the field functions both as the basic unit of education, training, and employment, and as the dominant reputational system, the variability and autonomy of sub-units is reduced. Different sub-fields in this situation have to be more oriented to one another than to outside groups if only to plan curricula and agree on the allocation of resources, at least temporarily. In most cases it is the disciplinary identity which is more important to practitioners than their, often short-lived, adherence to a particular topic or object of concern. Thus reputations in sub-units are usually complemented by attempts to gain the approval of scientists in other areas, especially when resources are scarce, through publishing in the more central journals or claiming more general significance for a study in a particular sub-field. Furthermore, sub-units are always competing for more resources and this usually means – unless jobs and funds are widely and easily available – demonstrating their importance for the discipline as a whole by redefining its goals and, perhaps, rewriting its intellectual history to boost a particular hero, as happens in sociology. Sub-group identity, then, is never quite enough on its own for most scientists as long as the academic discipline remains the central unit for recruiting, training, and allocating jobs and promotions through reputations.

Turning now to aspects of intellectual co-ordination and control, the heterogeneity of skills, audiences, and goals in fragmented adhocracies means that competition and conflicts range over a wide variety of issues and dimensions of cognitive structures. Individuals and groups diverge on many topics and preferences and there are few common standards which permit resolution of disputes. However, the wide availability of jobs and research resources – or their relative cheapness –

and the lack of a strong oligarchy controlling access to journals and books ensure that conflicts are not very intense nor long-lived. As long as individuals are able to obtain reputations which enable them to continue as researchers without having to convince large numbers of their colleagues of the importance and correctness of their work, they are unlikely to enter into sustained conflicts with others but rather to ignore conflicting views and continue to pursue their own paths. Thus diverse approaches to the description, interpretation, and explanation of phenomena will continue to be advocated and developed without major confrontations occurring between them or it being at all clear whether one is better than, or had overcome, others. Indeed, there may not be much internal co-ordination of research ideas and results between members of the same 'school' of thought, if they are geographically and organizationally separated and do not need one another to establish their competence.

The establishment and maintenance of research schools is, therefore critically dependent on the control system. If the employment structure is so diverse and differentiated that intellectual leadership is difficult to sustain over more than a relatively small group of students, there is little pressure to maintain intellectual conformity and cohesion within the group and members can pursue their own strategies in comparative independence of the leader. Given the lack of skill standardization and a formal, highly structured communication system, the ability of charismatic intellectuals to control the work of former students employed in a variety of organizations is limited. Instead, they may well be encouraged to publicize their differences with him or her as a way of demonstrating their own originality because of the wide availability of alternative audiences and resource centres. Rather than co-ordinating their research with one another, or combating the ideas and results of opponents, practitioners in these fields develop highly individual research strategies around distinct topics and problems often with idiosyncratic methods – or at least highly tacit and non-comparable ones – in order to obtain high reputations for originality. Differentiation of contributions is a higher priority here than co-ordination of results and contribution to the collective enterprise. This can become so extreme in some cases as to lead individuals to claim property rights over some empirical objects and try to exclude possible competitors from access to them. Thus anthropologists seem to avoid studying a society or tribe which has already been appropriated by someone else, and some sociologists of science claim expertise in particular sciences as a way of denigrating the competence of those who would challenge their results. Specialization here seems to be a strategy for avoiding communication with colleagues and integration of results in some general framework. The proliferation of case studies in the human sciences with the expansion of practitioners can be seen as part of this process of preferring differentiation and security to co-ordination and challenge.[28]

This sort of fragmentation of strategies and topics does not encourage theoretical work which aims to integrate and order a large number of contributions. Indeed, it promotes empirical diversity and differentiation around commonsense objects at the expense of integration and technical standardization. Despite the frequent claims to theoretical innovation in the human sciences, and the programmatic assertions of the importance of theoretical work, systematic integration of empirical work into a coherent theoretical structure is not very common in these fields and attempts in this direction are often dismissed as 'grand theory' or 'too abstract', or else remain restricted to critical analyses which rarely seem to generate further systematic research. This devaluation of theoretical co-ordination of tasks and strategies, most noticeable perhaps in management studies but also apparent in Anglo-Saxon sociology, political studies, and other human sciences, is linked to the specificity of cognitive objects and research problems. Typically, problems in these fields are not very general and abstract – Talcott Parsons's concern with the 'problem of order', for instance, was the exception in Anglo-Saxon sociology not the rule – and many of the systems and models being analysed in much human science are highly specific to particular common-sense objects rather than being exemplary for a large class of phenomena. It is the specific qualities and attributes of phenomena in these fields which are regarded as important – rather than their general significance and extension. Detailed facts are more highly valued than locating phenomena in some general scheme or exemplifying a conceptual problem. The resistance of some Anglo-Saxon social scientists to 'Continental' theorizing also reflects a concern with the specificity of cognitive objects and a reluctance to co-ordinate results through a common theoretical structure. Equally, although British sociologists do engage in abstract discussions and debates and write books about them for teaching purposes, there are remarkably few attempts at developing theoretical frameworks which integrate strategies and task outcomes systematically.[29] The importance of the undergraduate audience and the openness of the field to external audiences and goals[30] in this case limits mutual dependence among practitioners and encourages interpretative work on other thinkers rather than developing ideas on general problems which would lead to coherent research programmes.[31]

In summary, the high degree of technical and strategic task uncertainty in these fields restricts collective reputational control over individual research goals and procedures by formal, impersonal means and increases the importance of local exigencies and conditions of employment. Personal control and organizational structures have more impact on strategies and evaluations here than in sciences where central reputational control is established. Additionally, the plurality of audiences, of organizational goals and employment opportunities, enable scientists in these fields to pursue diverse goals and reduces their dependence upon one another and intellectual leaders. They are therefore able to orient their work to a wide variety of groups and purposes with an equal variety of research methods and procedures. There is little need to co-ordinate their research either within or between employment units and, in general, practitioners have a high degree of autonomy from reputational groups and research directors. Intellectual fragmentation is therefore a dominant feature.

(b) Polycentric oligarchies. In polycentric oligarchies the degree of task uncertainty remains high, but scientists are more dependent upon particular groups for reputations. Relative to fragmented adhocracies, this means that distinct research schools are likely to form, theoretical co-ordination becomes more important, and the level of conflict grows, as shown in Table 5.3. Such increases in the level of mutual dependence can arise through a growth in the number of practitioners relative to available permanent posts in an employment system oriented primarly to reputational goals and standards. The nineteenth- and twentieth-century German university system provides many examples of this process in a variety of scientific fields.[32] Competition for posts tied to reputations in a particular field here enhances reputational control over research strategies and so orients practitioners more strongly to collective goals and procedures than in fragmented adhocracies. Controllers of jobs and research institutes are able to control the direction and co-ordination of scientific work through the reputational system and so constitute a distinct 'establishment'.[33]

This control is limited in polycentric oligarchies, though, by the high degree of uncertainty about results, their interpretation and importance. The relatively low degree of skill standardization and formalization of the symbol system for communicating results means that research conducted in different centres cannot be compared and evaluated simply through the formal communication system but requires personal contact and relatively diffuse means of communication for its co-ordination. Thus, local contingencies and knowledge remain important for the integration and comparison of task outcomes. These factors prevent the development of a high degree of segmentation; instead they encourage the development of distinct research traditions and programmes around personal control of jobs and facilities. The importance of personal contact and knowledge for integrating research results and projects coupled with the need to demonstrate their general significance encourages the formation of sub-units around different approaches to intellectual goals. Because the plurality and equivalence of audiences and employers' goals is lower here than in the previous case, scientists have to co-ordinate their work more with one another but the high level of technical uncertainty prevents the formation of specialist sub-groups through the impersonal communication system. Thus distinct, strongly bounded sub-groups form through largely personal networks based on training and employment structures. Because of the difficulties of comparing results across these groups, and resolving disputes among them, these fields often manifest sectarian conflicts and become vertically differentiated by methodological and strategic preferences. Not infrequently, these groups publish in their own separate 'house' journals rather than submit results to discipline wide media.[34] These schools can therefore become mutually exclusive and sometimes bitterly opposed if resources are scarce and there is no general framework shared between them.

The differentiation of these scientific fields into distinct 'schools' of research controlled by leaders of employment organizations encourages some specialization of skills and problems within them because co-ordination can take place through personal contact and control. However, the highly tacit and diffuse nature of the interpretative skills needed to make sense of phenomena restricts this process. Furthermore, where appointments are tied to teaching posts which are not highly specialized, scientists have to be able to cover a wide range of material and be proficient in general knowledge. This, in part, explains the highly theoretical nature of German psychology when it was tied to chairs in philosophy institutes which involved teaching the history of philosophical thought.[35] Here again we see the importance of local employment conditions for the dominant styles of research in fields with relatively high task uncertainty.

The major influence of local leaders on work goals and methods is echoed in the differentiation of training programmes and the difficulty of transferring skills across research groups. Where the dominant approach to problems is co-ordinated with the selection and formulation of problems and work procedures by direct personal control, and training programmes are integrated with research in employment units, it is improbable that they will be uniform throughout the field or produce skills which can be directly applied to the full range of topics and concerns. Rather, distinct ways of doing research will be inculcated in the different centres and these lead to different ways of assessing results and competence. Wundt, for example, claimed that students trained in psychology at Leipzig would fail an examination conducted by someone trained at Würzburg.[36] However, some common elements are present in these fields, such as the insistence on psychology being an empirical, laboratory-based science, even if these mean different things in different places.

The importance of employment and training units in the formation and continuation of sub-units in polycentric oligarchies results in their relative prestige being strongly related to – if not dependent on – the prestige of the employing organization associated with them. Where this is high relative to other centres it will be easy for the school to dominate others while approaches based on relatively low status institutions will have difficulty in gaining attention in other centres of research. Such domination, though, is restricted by communication problems and the plurality of employment organizations in these sorts of sciences so that no single group can determine the goals and procedures of all scientists in the discipline. The extent to which schools are organized into a single hierarchy of significance and importance is therefore rather limited. While one department or institute may be able to place more of its graduates in chairs than its competitors, such as some at Oxford and Cambridge universities, and thus institutionalize its style of research as the predominant one, its control will never be total as long as other groups can control jobs and access to journals or books through other employment units. Furthermore, once established in a separate department, fields may wish to differentiate themselves from the school leader(s) and develop a different approach to disciplinary goals as Evans-Pritchard did in 1946 according to Kuper.[37] Thus British anthropology moved from being dominated by Malinowski and then Radcliffe-Brown in the inter-war years to being differentiated into a number of distinct schools located in Oxford, London, Cambridge, and Manchester built around powerful professors who collectively dominated appointments, research funds, and publication media in the 1940s.[38] This group of professors effectively operated as a cartel of equals and no single school dominated the field as much as Radcliffe-Brown had done before the war when the number of jobs was much less.

In addition to the prestige of sub-groups being roughly

equivalent in these fields, their internal structure tends to be more similar than in fragmented adhocracies. Because in any one national system these tend to be fairly similar, knowledge production and validation sub-units within a nation state operate in a similar way and have similar patterns of authority. In the cases of German and many British university departments, for instance, the professor dominated most aspects of the research programmes and the intellectual structure was hierarchical with each member being given a particular problem or entity to study within the overall frame work of the schools. Functional dependence within each group was, therefore, relatively high while the common theoretical structure reduced intellectual uncertainty to some extent.

The need to integrate results around common intellectual goals within each school and to demonstrate their general importance mean that theoretical co-ordination is accorded a greater value in these fields than in fragmented adhocracies. Even if the dominant tone is empiricist – as is often the case in Anglo-Saxon human sciences – [39] the need to fit one's results into the dominant orthodoxy ensures that theoretical relevance is an important concern and the disputes between opposed schools heighten awareness of theoretical issues in the field as a whole. Review articles which bring together a number of studies and how they fit together around the main tenets of the schools are an important means of strengthening the theoretical framework and extending it so that group leaders periodically use this format to synthesize the school's achievements. Co-ordinating research through the group's framework is therefore an important activity which is often reserved by the leader, thus heightening the status of theoretical synthesis.

Given the divergence of theoretical and procedural approaches between the various schools in these fields, and their relatively high strategic dependence, it is not surprising that intellectual conflicts tend to be both wide in scope and intense in degree. Despite their common tutelage by either Malinowski or Radcliffe-Brown – or perhaps because of it – British anthropologists engaged in vitriolic and highly personal disputes in their attempts to differentiate themselves from one another and claim reputational hegemony. Also, disputes in early North American anthropology culminated in the dismissal of Boas as a member of the American Anthropological Association's governing council and threats of expulsion from the Association in 1919.[40]

The importance of theoretical integration and relevance as criteria for awarding reputations is reflected in attempts to theorize cognitive objects in polycentric oligarchies. While commonsense descriptions and concepts may still predominate in the characterization of intellectual concerns, phenomena are here also described in more technical terms and ordered in some theoretical structure. Thus detailed accounts of empirical objects are accompanied by claims to theoretical relevance and generality. Particular facets are highlighted and emphasized according to the theory being used and the culture, tribe or personality are seen as exemplifying more general and abstract phenomena than in fragmented adhocracies.

(d) Partitioned bureaucracies. I suggested in Chapter 4 that reputational organizations combining a low degree of technical control of empirical phenomena with a high degree of uniformity and stability of probelms and approaches were unlikely to become established and remain stable unless they partitioned the central core of conceptual orthodoxy away from emprical sources of uncertainty[41] and yet managed to retain organizational control of the empirical areas. The overall level of dependence upon the reputational system has to be high for these sorts of fields to maintain a relatively low degree of strategic task uncertainty and the dominant theoretical approach. Despite, then, the development of many 'applied' sub-fields based on relatively everyday and common-sense objects such as 'art', 'education', and 'labour' in which the basic framework is simply brought to bear upon empirical phenomena, these do not become established as semi-autonomous reputational organizations but remain subservient to the overall reputational system which values analytical elaboration and development more highly than empirical exploration. The degree of segmentation around common-sense objects is therefore limited by the domination of theoretical goals.

Within the central core analytical skills are highly standardized and the symbol system is also highly structured so that central reputational control over research strategies and performance standards can be maintained through the formal communication system. The relatively low degree of strategic task uncertainty means that scientists can specialize in analytical refinements to the dominant framework and do not conflict over a wide range of substantive issues. The scope of competition for reputations within this central core is restricted to technical refinements and extensions of the analytical scheme scheme to novel problems which can be tackled with existing techniques. Given the high degree of mutual dependence in this core, the intensity of competition is considerable but limited to increasing the coherence and completeness of the theoretical framework. High reputations for significant work are based on theoretical sophistication and demonstrating how the dominant approach is consistent and general, so both objects and problems are very general and abstract in the core area.[42]

The strong separation between the central core area and the peripheral sub-units in this type of science makes any simple characterization of the field as a whole difficult to achieve since what is true of the core need not be so of all parts of the field. In particular, the degree of impersonality and formalization of co-ordination and control procedures varies between sub-units as the relatively intractable and unpredictable nature of empirical phenomena restricts the extent to which systematic comparison and validation of task outcomes can be accomplished through the formal communication system. The relative openness and lack of standardization of empirical descriptions – linked to the high extent of technical task uncertainty – limits the integration of results into the theoretical structure and so encourages the partitioning of empirically based sub-units from the central core. These aspects also restrict the generality of problems and objects in the peripheral areas as scientists have to develop a greater understanding of the specific details of phenomena and limit the degree of abstraction in their analyses if they are to obtain coherent results. In the central core, on the other hand, theoretical integration and abstraction are at a premium as the dominant framework is elaborated and refined. Given its centrality this means that the science as a whole manifests a highly abstract character with a highly standardized and formalized symbol structure for communicating ideas and results but this does not imply that all areas are equally highly formalized. Heterogeneity of sub-units is thus high.

Control over the field as a whole is achieved by a high degree of standardization of analytical skills and rigid delimitation of the sorts of problems which are legitimate and important. As long as the reputational system is able to control relatively prestigious reputations for research on these problems the failure of the standard skills to deal with related but non-uniform problems in peripheral areas will not be sufficient to reduce their legitimacy and prestige. Because the skills and techniques are relatively successful in dealing with the restricted, theoretically enclosed problems at the centre of the subject, deviations at the periphery can be largely ignored by the intellectual élite who control the dominant communication media and training programmes. Since analytical skills provide the basis for practitioners' distinct and socially valued identity, they are unlikely to challenge their utility and adequacy when they demonstrably fail to solve less restricted problems but rather seek to translate these issues into terms which enable their techniques to be applied or else reject such problems as being irrelevant or unimportant. Thus if people do not behave 'rationally', i.e. according to the tenets of the dominant theory, this demonstrates either their incompetence or the irrelevance of their behaviour rather than any important flaw in the theoretical apparatus.

Many, if not most, of these features of partitioned bureacracies are found in post-1870 Anglo-Saxon economics. A number of observers have commented upon the disjunction between theoretical analysis and empirical research, on the difficulties of arriving at shared interpretations of task outcomes in empirical studies, and the lack of reliable, visible, and stable results in empirical work in economics.[43] Additionally, while some analysis see variations in approaches between competing 'schools', others consider such divergencies to be fairly minor and little change observable in the dominant theoretical framework since 1870, if not since Adam

Smith's *The Wealth of Nations*.[44] This framework seems well entrenched in the major graduate schools, the central journals, and the international prestige system. Tight control is exercised over intellectual priorities, the selection of 'real' problems, and the sort of analysis which is admissible in contributions to economic knowledge.[45] Of especial note in this respect is the domination of undergraduate and post-graduate instruction in economics by a small number of textbooks which inculcate a distinct and rather rigid set of intellectual practices.[46]

The importance, coherence, and similarity of textbooks in economics are perhaps most similar to the role and character-istics of textbooks in some natural sciences as described by Kuhn.[47] The highly restricted nature of phenomena in economics, at least since the marginalist 'revolution',[48] has enabled the field to be presented to recruits in a systematic, closed manner which encouraged values of coherence, sim-plicity, and formalism over those of accuracy, applicability, and empirical relevance. Just as Kuhn described the learning process in physics as 'a dogmatic initiation in a pre-established tradition that the student is not equipped to evaluate'[49] and emphaized the crucial role of concrete problem solutions in producing closed mental sets, so too economics textbooks present the field as fixed body of doctrines and 'laws' and often give students a number of highly abstract and general 'problems' to work through as a means of developing competence. A degree in economics thus implies a capacity to solve artifical analytical problems and not an ability to understand everyday events and phenomena in the economy.[50]

The consequences of this rigorous training in solving problems in economic theory are a high degree of uniformity in the analytical skills and outlook of economists, strong consciousness of the boundaries of economics and what are, and are not, economic problems, a highly rule-governed set of research practices which are strongly oriented to theoretical and analytical goals and a highly formal symbol system for communicating and co-ordinating task outcomes.[51] However, coupled with this coherent body of doctrine and practices in theoretical economics is considerable confusion and disputes about its relationship to economic 'realities', as witnessed by the seemingly interminable methodological debates in the 'theory of the firm' and macro-economics.[52] The price paid by the insistence on analytical coherence and restriction has been the increasing difficulty of using economic theory to explain empirical phenomena. As Jenkin has observed,[53] economics does not have a tradition of testing its theories in a comparable manner to physics, although in many other respects the two fields are quite similar as can be seen from Table 5.3, and there are no standardized means by which empirical studies can feed back into the theoretical corpus. Instead, what happens is that the analytical structure is 'applied' to a wide variety of topics and problems by redescribing these in terms of concepts and categories taken from theoretical economics and 'solving' them in the ways taught in undergraduate textbooks. The results of these 'applications' do not make much impact on the dominant theoretical structures precisely because they are simply applications. Thus numerous topic areas are formed around such problems but do not function in the same ways as sub-fields of physics do. Their interdependence is less and their result rarely affect work in other areas. Because the output of these areas is not oriented to theoretical goals, and thus is not significant for 'real', i.e. analytical, economics,[54] it does not lead to high academic reputations and makes little impact on the dominant doctrine which continues to be reproduced without substantial modification even when contradictions and errors are admitted.[55]

The major division in economics, then, is between the theoretical, analytical core which dominates training programmes, the communication system and academic reputations, posts and honours, and the applied periphery of problem fields which take their methods and concepts from the core but do not contribute to its modification or improvement. Within analytical economics there is a high degree of mutual dependence among practitioners, a high degree of formalization and standardization of work procedures, assessment standards and problem formulations, and a high degree of task differentiation. In the applied areas there is often a similarly high degree of mutual dependence among academic practitioners but the existence of other audiences and resources controllers obviously mitigates this in many cases. Difficulties in applying and extending the dominant doctrine to empirical matters reduces the degree of formalization and standardization of research practices and thus co-ordination of results is not so formal and impersonal here. Because control over work goals is less monolithic in the applied areas, there is some scope for some theoretical deviance as in industrial, labour, and welfare economics.[56] Similarly, the extent and degree of task differentiation is lower here as originality can be demonstrated by novel applications and interpretations of empirical material as well as by additional theoretical refinements. The high degree of analytical coherence, stability, and uniformity in the central core of the field is thus accompanied by a diversity of peripheral areas of application which share the same basic training and commitments but otherwise have little connection with one another or with the central core. This structure is capable of absorbing considerable increases in the number of personnel and other resources without undergoing fundamental change by simply expanding the range of 'applications', to, for instance, the economics of education or the economics of the arts, or by increasing the degree of formalization and differentiation at the centre. Co-ordination problems are 'solved' by the extension and intensification of standardized doctrines, problems, and research practices in the central core and separating it from the peripheral areas.

No other field in the social sciences and humanities comes close to economics in its combination of theoretical uniformity and technical uncertainty. Attempts at formalizing and standardizing fields have usually failed because of the difficulties of establishing a highly restricted understanding[57] of their subject matters and/or the plurality of possible audiences and resource providers which has limited the degree of dependence of scientists upon one another for reputations and material rewards. Generally, the closeness of most areas to commonsense concerns and concepts has severely hampered movements to 'professionalize' them and render lay judgements totally irrelevant to reputations. The separation and integration of theoretical work around a small number of basic axioms and assumptions have not developed in most human sciences because of the availability of alternative conceptions of their subject matters and of groups to legitimate them. The 'poly-paradigmatic' nature of many social sciences, noted by Martins and Lammers,[58] seems to me largely due to this plurality of audiences and the importance of everyday concerns in formulating and selecting problems. As long as technical control is limited and mutual dependence among practitioners restricted by the 'educated public' being a legitimate addressee of task outcomes, the development of a highly formal theoretical structure with standardized work procedures is improbable. Where groups have tried to formalize techiques and restrict admissible problems, as in causal modelling in sociology for instance, they have successfuly established themselves as a legitimate sub-field within access to journals and jobs but have not dominated and unified the whole field.[59] Many fields exhibit a mixture of high and low formalization and standardization in their constituent sub-areas without any single one becoming dominant. Given the generally high degree of technical task uncertainty and hence personal means of work co-ordination and control, this is not surprising when the overall size of the field is fairly large and resources are not highly concentrated as is largely the case since the expansion of university posts in the 1960s.

(e) Professional adhocracies.   When technical task uncertainty is lower than in the examples discussed so far in this chapter, task outcomes can be compared and evaluated across work places through relatively formalized and esoteric symbol systems. This enables research to be co-ordinated through the communication system, and problems to be divided into separate tasks and skills across employment organizations, thus increasing the degree of functional dependence and of segmentation. Reputations can therefore be obtained on a more impersonal, general, and cosmopolitan basis from the field as a whole. Also, the size of the organization can be larger in these fields without fragmentation occurring because personal contacts are not so crucial for interpreting and connecting results. The greater standardization of research

procedures and training programmes implies stronger organizational consciousness and identities than in cases *(a)* and *(b)* and a corresponding emphasis on following collective rules correctly. The technical details of research in professional adhocracies are more esoteric and removed from commonsense understanding and concepts than previously and 'professional' skills are largely defined, controlled, and evaluated by practitioners.

Problems are, however, varied and frequently changing. The theoretical environment of scientists in professional adhocracies is quite uncertain and goals are diverse. Whereas control over research methods and processes is exercised by the field as a collective totality, the choice, formulation, and relative importance of problems is more subject to individual, employer, and state control. Diversity of research programmes and aims is thus encouraged and overall integration of theoretical objectives limited. These sorts of scientific fields are segmented into numerous problem areas which vary in their degree of interconnectedness and interdependence as well as their degree of stability. Sub-unit boundaries are fairly fluid and permeable as specialized skills are used for a variety of problems and goals.

In professional adhocracies, then, control over research goals and strategies is often shared between reputational groups, employers, and funding agencies. While skills are standardized and task outcomes fairly clear and unambiguous, the need to co-ordinate problems and approaches with particular collegiate groups is limited so that scientists pursue a wide variety of topics and goals. Reputations can be gained here from a variety of audiences and sub-groups in and outside the field without having to demonstrate the general significance of one's contributions to a particular reputational group. Thus numerous sub-fields can form and award reputations without any strong need to integrate and co-ordinate results and problem areas. Their interdependence is low and precise interrelationships between results in different areas difficult to establish. So, no one sub-field dominates the whole reputational organization, and where integration of diverse problem areas is attempted this is as likely to be undertaken by employers or funding agencies for non-intellectual goals as by scientists pursuing reputational objectives. In such instances, the means employed are as often financial and managerial as theoretical. Local influences on research strategies and priorities are, therefore, considerable and problem areas are not organized into a stable hierarchy of intellectual significance.

Competition in these fields is not very intense because resources are relatively widely available and strategic dependence is low. Since reputations can be obtained from specialist groups without convincing the whole professional group of the significance of particular problems and results, differentiation of topics and techniques is encouraged and conflicts over strategies limited. Additionally, since scientists do not need to show how particular approaches and problems fit in and affect general issues, theoretical co-ordination of work across subfields is not very important here and theoretical conflicts not a major phenomenon. However, because strategies do differ and problems vary in their perceived importance, the scope of conflicts is potentially considerable although not as great as in a field with both high technical and strategic task uncertainty.

Co-ordination and control of task outcomes within specialist groups is achieved through standardized skills and procedures. However these do not uniquely determine the sorts of problems tackled, nor their relative importance, and significance standards are pluralistic and often unstable. Control over work goals tends therefore to be heterarchical rather than hierarchical. This is exemplified by many bio-medical fields where employment is often in non-academic organizations whose goals are not identical to those of university institutes and departments. However, despite the local, fluid, and contextual nature of much research in laboratories described by Knorr, Latour, and Woolgar and others in these fields,[60] the level of technical control and standardization is greater than that found in the human sciences. This is shown in the growing standardization and mechanization of analytical techniques.[61] Similarly, common research procedures and methods are taught in training programmes in different universities even if they vary in departmental labels and boundaries.[62] Biochemistry, for instance, has been institutionalized in different ways in different national education systems, but a modern biochemist has a similar set of skills and practices whether he or she was trained in the USA, England, or the Netherlands.[63]

Where the bio-medical fields differ most from traditional academic disciplines is their combination of different skills and practices to work on problems which do not fit clearly within disciplinary domains. Thus research methods are fairly formalized and homogeneous across research sites, so that results can be compared and assessed for their mutual implications, but their particular combination in the pursuit of a particular problem is less standard and more open to local contingencies than in, say, mathematics. It is this particularism which renders reputational units varied and fluid so that sub-areas are not easy to identify and scientists seem to have plural and diffuse identities. Furthermore, their goals vary in their generality as do the models studied. Biologists, for instance, may have quite general goals in cancer research but often focus on a single system in great detail while biochemists analyse a range of substances but have a more specific and limited objectives.[64].

Specialization in these fields is largely a matter of topic or problem, but is also affected by the particular way in which skills have been brought together for a particular purpose and thus more by the structure of the major employing organizations in the field than occurs in more academically dominated disciplines. In the extreme case, decisions by a relatively small number of research directors in, say, cancer research, can lead to the formation of a new reputational sub-group which is located completely outside universities.

The consequences of this variability in influences and research practices are a high degree of specialization and differentiation of groups and topics with considerable fluidity of membership and boundaries. Audiences are diverse and shifting with a variety of interests and purposes. Reputations are equally fluid and unstable and an individual's dependence on any one group of colleagues is low. Because of the closeness of medical interests and goals, at least for public legitimation and fund raising, scientists pursue diverse goals and strategies without integrating results across specialist concerns or, often, across an individual's own output.[65] Co-ordination of work is conducted more by employment organizations than by stable cross-organization collectivities and more local variation in goals and problems tackled occurs here than in, say, mathematics. The degree of segmentation, although greater than when technical task uncertainty is high, is thus limited in professional adhocracies. Individual autonomy from a single specialist group of colleagues is quite considerable but dependence on employer's priorities and resources is often substantial as the Latour and Woolgar study shows.[66] Control over research goals is thus shared between the individual, the employment organization, and a number of rather fluid and unstable reputational groups. Their co-ordination is often effected by employers and funding agencies' programmes rather than through any single group's journals, and overall theoretical synthesis of results is infrequent. Individual research strategies tend to be rather *ad hoc* and short term as is to be expected in the highly uncertain intellectual and organizational environment.

A similar diversity of audiences and goals but with a stronger core of technical expertise and skills is found in artificial intelligence.[67] Here a common reliance on large computers and elaborate programming skills provides the basis for organizational boundaries and identity which are reproduced through international conferences and journals. However the varied goals pursued, and frequent changes in them, mean that distinct sub-groups from around different problem and topics which sometimes, as in the case of theorem proving, become autonomous and separate from the main field. The growing commercial relevance of much of artificial intelligence is likely to encourage this process of differentiation by problem area, but as long as the craft skills remain coherent and controlled by the reputational leaders the frequently announced dissolution of the field is improbable.[68] Sharing control over work goals, that is, leads to problem diversity and fluidity of reputations without necessarily creating intellectual and social fragmentation where skills and procedures are sufficiently coherent.

Mathematics in the USA as described by Hagstrom, Fisher, and Hargens[69] seems to be an example of a scientific field where low technical task uncertainty is combined with high strategic task uncertainty but with a greater degree of strategic dependence than the previous two cases. The high degree of formalization and standardization of work procedures and methods is maintained by the common undergraduate and early postgraduate training programmes in universities combined with the academic domination of the job market. The discipline of mathematics remains the major unit of inculcating skills and providing employment despite the growth of applied mathematics and industrial employment.[70] However, the general availability of jobs in the post-Sputnik era and the cheapness of research materials means that, once certified, practitioners are free to pursue a wide variety of problems without worrying too much about their overall significance for the discipline as a whole. As long as they are competently carried out and follow collective rules and procedures, research projects on a broad range of topics should lead to publications and reputations. Originality is demonstrated by applying and extending techniques to cover more esoteric and specialized problems and so the field becomes highly internally differentiated by sub-groups following diverse and distinct goals with little need for their co-ordination and integration. Hagstrom has qualified this state of affairs as 'anomic' because of the lack of attention mathematicians seem to give to their colleagues' work and hence the breakdown of his recognition-exchange model of the scientific community.[71] However, he ignores the continued existence of some central problems for mathematics as a whole and the crucial role of the journals in structuring communication and hence knowledge. Furthermore, the diversity and differentiation of North American mathematics is controlled by the fairly standardized training system and doesn't seem to have stopped mathematicians seeking reputations or making contributions to collective goals. The search for autonomy and independence has resulted in a high degree of specialization and limited the extent of integration and synthesis of specialized contributions but all this means is that the dominant research style and nature of task outcomes is different than in other fields. It does not imply a total breakdown of social and cognitive order and there are no obvious reasons why this prestigious professional group is about to commit collective suicide.

(f) Polycentric professions. If the degree of strategic dependence is greater than in professional adhocracies but levels of technical and strategic task uncertainty remain similar, three major differences in their internal structure are noticeable. As shown in Table 5.3, polycentric professions are more likely to develop distinct research schools, to emphasize the theoretical co-ordination of research and intellectual strategies, and to exhibit a greater degree of competition between rival groups than are professional adhocracies. Compared to polycentric oligarchies they also exhibit more specialization and standardization of tasks and skills, rather more tendency to form international specialisms around particular aspects of general goals, and a greater formalization of co-ordination and control procedures because of their lower level of technical task uncertainty. The relatively high degree of strategic dependence means that scientists have to show how their work 'fits in' and makes a difference to the work of others. Pursuing specialized and distinct problems without much regard for their general significance for collective goals is not so feasible in these fields since reputations depend more upon the collegiate group and less upon a wide diversity of audiences. Consequently a greater awareness of theoretical implications and concern to demonstrate the overall importance of particular problems and topics is to be expected here.

This implies that the reputational organization has greater control over work goals and strategies than in the previous instance. If dependence upon the collegiate group is greater then the influence of employers and funding agencies on goals is reduced. However, this reduction is only relative and does not imply a lack of local influence. The existence of theoretical pluralism and variety in these fields means that different approaches and priorities are often associated with different employment organizations and training programmes. Funding agencies can also promote a particular strategy or set of priorities, as the Rockefeller foundation and the Carnegie Institution did in biology for example,[72] but this pluralism and diversity are controlled by the reputational system as a whole evaluating the general fruitfulness of different approaches. Separate problems and research styles need to be justified in terms of their overall contribution to the work and goals of colleagues rather than being followed in relative isolation. Thus scientists have to orient themselves to what a wide range of colleagues are doing and seek to influence a considerable number of them if they are to obtain major reputations.

Problem areas are therefore not likely to become so separated and differentiated in these sorts of scientific fields as in the previous case. The degree of interdependence of problems and approaches is greater and their relevance to the central questions of the field more critical to scientists' overall reputations. Technical standardization enables specialization to develop across research sites and co-ordination to take place through the formal communication system so that sub-groups can form around particular problems, but their mutual relationships and importance *vis-à-vis* the field as a whole are of considerable concern so that sub-fields compete for centrality and the ability to determine collective priorities.

The need to demonstrate general significance and relevance encourages close links between general theoretical and methodological approaches and the formulation and implementation of research strategies. The choice and characterization of problems therefore become tied to different perspectives about the nature of the field and appropriate ways of proceeding. Priorities and criteria for the assessment of the significance of particular problems differ between groups in these fields so that theoretical diversity is associated with problem diversity. Thus the differentiation of groups and activities will be more determined by conceptual and programmatic differences than by the particular objects or systems studied. Conflicts and disputes over the appropriateness and validity of competing approaches are therefore frequent, as groups insist on the importance and correctness of their goals and strategies. Generally, the rise of molecular biology can be seen as a successful attempt to redirect the goals and strategies of the biological sciences which involved a change in the nature of the problems considered important and how they were to be tackled.[73] Theoretical priorities and views were intimately connected to the sorts of problems which were considered crucial and to the development of new techniques and work methods. Because this change involved altering research procedures and practices, and the relevance of particular skills, it is not surprising it needed sustained financial support and took a considerable time to accomplish.

The emergence of one research programme and school as dominant in polycentric professions is likely to be strongly resisted as long as competing programmes have access to resources and can claim success in terms of the standard work procedures and validation critieria. A stable hierarchy of sub-units based on distinct problem areas is therefore improbable unless the whole basis of the science is radically altered and new criteria of fruitfulness and significance are established in which case the field effectively becomes transformed, as physiology has been by biochemistry and molecular biology.[74] Instead, competing programmes focus on different problems and 'models' – as with Michael Foster's focus on the heartbeat and myogenic material[75] – in the light of their own significance criteria and programmatic preferences. Each research school is able to develop its distinct programme through its control of jobs and facilities, and often its own journal, while engaging in debate and disputes with competitors. Foster, for example, was able to establish the Cambridge school of physiology through his control of the *Journal of Physiology* and his post as Biological Secretary to the Royal Society, as well as being able to rely on the private means of his students and collaborators.[76] Without these resources it is doubtful if his school would would ever have made the impact it did upon the field of physiology, or even existed as a distinct unit of knowledge production over more than a single generation.[77]

Competition between schools in these scientific fields heightens awareness of theoretical differences and encourages theoretical integration of results and strategies within them. Thus the intensity of intellectual conflicts in the field is quite considerable especially if the degree of strategic dependence is very high, while orthodoxy inside each group is strongly controlled through a central intellectual leader. Because of the relatively standardized technical procedures and symbol structres, research can be fairly specialized but the overall theoretical implications of results are important and so both research problems and objects are selected with a view to their general importance and significance.

The scope of intellectual conflicts is not as great as in fields where technical task uncertainty is very high because of the increased standardization of skills. Biologists, for instance, did not dispute the validity of results in the conflicts between geneticists, embryologists and paleontologists in the early twentieth century to the same extent as some social scientists do, but rather disagreed about the significance of problems and the scientific status of different styles of research.[78] Research schools in the sense discussed by Geison[79] are not, then, the same as conflicting groups in some of the human sciences because the former produce more specific knowledge with more standardized procedures which enable results from different work-places to be compared and co-ordinated. Thus schools in the biological sciences are not so dependent upon personal contact and local continuity for their work to be effectively co-ordinated and controlled. The meaning of results is relatively easy to establish and communicate through the formal communiation system in these fields. Conflict therefore focuses more at the level of research strategies and significance of problems, and is generally more specific than in fields with a high degree of both technical and strategic task uncertainty.

The high degree of functional and strategic dependence heightens collective consciousness and an awareness of common concerns and rules. Even though it also increases mutual competition for reputations, and so emphasizes conceptual and theoretical differences, greater dependence focuses attention on a particular collegiate group and the need to influence their priorities and strategies. It thus reduces tendencies towards fragmentation and strengthens organizational boundaries. The standardization and specificity of skills also reinforce common identities and exclude 'amateurs' and outsiders from participation in debates and research. However strong conflicts may become within the field, the high degree of mutual dependence implies considerable organizational solidarity and cohesion. Disputes over strategic priori-

ties and the significance of problems and hence of contributions and reputations, rarely transgress organizational boundaries in these fields because reputations are controlled by the
group as a whole and these are sufficiently prestigious for
practitioners not to seek outside audiences. These sorts of
fields are therefore not likely to develop in sciences which are
not central to currently dominant ideals and /or which have
low social prestige.

(g) Technologically integrated bureaucracies.   In fields with lower
degrees of task uncertainty, work can be more systematically
planned and co-ordinated. Division of tasks and skills can be
quite extensive since task outcomes are fairly predictable and
reliable. Problems can be divided into separate aspects for
different people to work on both within and between
work-places so that the scope of an individual scientists's
concerns is quite limited. Skills are standardized across
training programmes and are firmly bounded. Additionally,
the selection and formulation of research questions involves
fewer uncertainties than in the previous fields so that
reputations are more stable and predictable. What important
problems are, and how they should be tackled, is fairly clear
and widely accepted. The theoretical implications, and
general significance, of research results are relatively unambiguous and not likely to be the focus of major conflicts or
subject to sudden changes.

Standardization of tasks and skills in fields with relatively
low degrees of technical and strategic uncertainty reduces the
level of local variability of research practices and interpretations. The degree of segmentation of the field into separate,
yet commensurate, problem areas which cross employment
units is therefore quite considerable while the establishment of
distinct schools pursuing opposed goals and priorities is
improbable. The highly impersonal and formal communication and control system enables work in many different
research sites and countries to be connected and co-ordinated
within each specialist area which form the major organizational unit for the day-to-day management of research. Thus
the strongly rule-governed nature of intellectual work in
sciences with relatively limited degrees of task uncertainty

enables a considerable degree of decentralization of operational control to specialist groups in a similar way to that in other work organizations.[80] However, differences in the degree of strategic dependence between scientists and groups are associated with variations in the organization of work in these fields, as summarized in Table 5.3.

Variations in the degree of strategic dependence in fields with low task uncertainty affect the extent to which practitioners have to co-ordinate their research with that of other groups throughout the field and influence general priorities to gain high reputations. Where the environment is relatively benign so that resources are generally available and practitioners can specialize in specific topics without worrying too much about their general relevance to the field as a whole, i.e. where sub-groups control jobs, equipment, and journals in relative autonomy from the parent 'discipline', the organization can function as a decentralized bureaucracy. Because the general theoretical background is fairly stable and appropriate problems clearly specified, interrelationships between problems and sub-fields are not difficult to discern. Where the degree of strategic dependence is relatively low, this means that severe disputes are not likely to arise because reputations in different problem areas do not need to be mutually ordered into some sort of priority for continued access to resources.

The internal structure of scientific fields with relatively low degrees of task uncertainty and strategic dependence is highly differentiated as practitioners seek to differentiate themselves from one another by studying distinct topics or using different analytical techniques. Because the general theoretical framework is stable and not in substantive dispute, the significance of each specialized contribution is fairly clear and does not require sustained debate. Consequently, scientists are encouraged to pursue highly specialized topics and deal with narrowly defined problems. Co-ordination of results increasingly takes place through the technical apparatus used to standardize the raw materials and transform them so that the sorts of problems which can be investigated are circumscribed by this apparatus and its presuppositions. This ensures their uniformity and stability as well as standardizing their interconnections and mutual implications. As highlighted by

Bachelard in his discussion of 'phenomeno-technics',[81] theoretical structures are embodied in chemical techniques which realize chemical phenomena so that every experiment that uses a particular set of techniques is automatically interrelated with other uses of those procedures and the theoretical significance of the results is given by the analytical devices used. This means that theoretical integration of results can be accomplished without extensive personal contact and debate through the formal communication system and knowledge accumulates in a rule-governed manner. Problems fit in with one another 'naturally' without much need for negotiation or theoretical elaboration.

Sub-units develop around specialized problem areas in these sciences, and become increasingly differentiated as numbers expand and facilities become generally available in work-places. Because the degree of strategic dependence is fairly low, scientists do not have to orient their research to the entire field but rather can restrict themselves to obtaining reputations from a small specialist group which is probably able to control access to jobs and research funds. As long as this situation continues, there will be little need establish a hierarchy of sub-units in terms of their importance to the field as a whole and consequently to order problem in terms of their theoretical significance.

This, in turn, means that theoretical integration and co-ordination of problems is not especially valued in these fields and so theoretical work is not monopolized by a separate group of scientists who are able to dominate significance criteria and the ordering of reputations between sub-units.[82] Equally, although cognitive objects and problems are more general and theoretically formulated than in adhocratic fields because of the theoretically structured techiques, they are not very abstract and highly removed from empirical details. In fact, if we take Shinn's account of research in contemporary mineral chemistry as an instance of a sub-unit of a technologically integrated bureaucracy,[83] reliance on specific qualities of objects seems still quite high and there is a considerable emphasis on detailed observation of basic characteristics. The particular features of individual substances and compounds are still important for chemists and their qualitative variations form a major basis for problem formulation and separation. As Georgescu-Roegen points out,[84] chemistry is not arithmomorphic in the same way that parts of physics are, and seeks to account for change and variation through general processes without reducing them to simple arithmetic continua.[85] The generality of objects and problems here is thus limited.

The high degree of skill standardization and low degree of technical and strategic task uncertainty ensure that the field as a reputational organization exercises considerable control over research strategies and procedures in the public sciences. Local autonomy is correspondingly reduced for practitioners seeking reputations for competent and significant work in the field or a sub-unit of it. By and large, academic departments undertake similar types of research and follow similar general strategies in seeking high reputations so that separate schools based on employment units do not from in the same way that they do in polycentric professions. Equally, national differences in research styles and approaches, such as the debates over the theory of solution in the late nineteenth and eartly twentieth centuries chronicled by Dolby,[86] are less apparent in contemporary chemistry. For the same reasons conflict is both limited in scope and in degree in these fields when strategic dependence is also low so that the overall significance of problems and results is not a focus of disputes.

Relative low strategic dependence also results in intellectual identities and boundaries being a matter of only limited concern. While skill standardization and theoretical orthodoxy do lead to a distinct professional identity and high reputational control over work procedures and evaluation of competence, internal demarcation of sub-groups and boundaries between the reputational, public science organization and other areas of chemical research are not as strongly emphasized and policed as in fields where strategic dependence is greater. The traditionally strong links between chemistry and industry, possibly since the eighteenth century[87] and certainly since the nineteenth century,[88] have limited the extent to which industrial and academic research have been separated and have provided additional sources of facilities and funds.

(h) Conceptually integrated bureaucracies.   When a low degree of task uncertainty is combined with high strategic dependence, this smoothly functioning bureaucracy becomes more competitive as scientists seek to demonstrate the importance of their work to a wider range of colleagues and claim greater significance for their problem or topic. Skills and techniques remain, of course, highly standardized and problems are still relatively uniform and stable, but the neçessity of convincing the reputational organization as a whole of the relevance of one's sub-field for general collective goals results in disputes over how these goals are to be interpreted and applied in particular instances. The relative priority of sub-goals and groups within the overall theoretical framework becomes a focus of conflicts. It is not the actual framework itself which is disputed but the way it is used to integrate and order reputations across sub-fields. Thus, where dependence on one's colleagues is high, because of restricted access to limited resources for instance, the overall importance of problems is more critical than in the previous case and how collective goals are interpreted has major consequences for individual and sub-group reputations. If their results and problems cannot easily 'fit in' with dominant objectives and exert distinct influence on others' research strategies, scientists are unlikely to gain sufficiently high reputations to continue making competent contributions. Since it is here more important to contribute to the field as a whole – as distinct from making useful but limited contributions to specialist groups – practitioners conflict over the relative priority of their various specialisms.

The greater need to co-ordinate sub-units in these fields means that the integration of problems and strategies in separate areas cannot be left entirely to standardized skills and apparatus. Because scientists in different areas disagree about the relative importance of their problems for the discipline as a whole, technical co-ordination of results is no longer sufficient for reputations to be awarded in an acceptable fashion. Thus, while co-ordination and control processes are impersonal and formal, and the scope of intellectual conflicts is small in view of the relatively low amount of uncertainty, disputes can be intense and involve general issues of disciplinary identity. The centrality of particular issues and problems to the field as a whole has such major implications for reputations, and hence continued access to valued resources, when strategic dependence is very high that scientists struggle over the precise relationship between areas and the generality of results. Although they work on highly specialized problems and form distinct sub-fields which exercise some control over reputations and resources, their mutual interconnections and implications are much important to practitioners here than in the previous case.

This means that sub-fields and problem areas are not so sharply distinguished and separate as in technologically integrated bureaucracies. The existence, and importance, of the unified theory which orders relations between sub-fields enables scientists to move more often, and more easily, between problems and topics than they do in other sciences.[89] This movement is facilitated by the high degree of generality of objects and of problems which such a unified theory provides and requires. Unlike chemistry, and most other sciences, the specific properties of particular phenomena are of limited interest in the most 'fundamental' parts of physics and detailed understandings of individual objects are not sought.[90] Rather, investigations are 'restricted' to a small number of highly general and abstract relations which can be expressed in coherent formalisms and 'applied' to a large number of phenomena.[91] Specialists in particular systems or problems are, then, able to transfer their interests and skills with less difficulty than in fields where detailed and specific knowledge of particular objects is more important. As can be seen from the 1972 National Academy of Sciences survey, physicists do indeed 'migrate' to a considerable degree between sub-fields with one-third of PhDs changing their major interests in 1968–70.[92] Reputations, then, are more discipline-based than in chemistry and sub-fields are less independent of disciplinary goals and ideals.

Increasing strategic dependence among practitioners and specialist groups generally leads to more leads to more emphasis being placed upon theoretical integration and co-ordination of results and problems as scientists struggle to impose their priorities and standards of significance upon their olleagues. However, only in physics, and especially nuclear physics, has this emphasis resulted in the establishment of a distinct occupational identity for theoretical work and hence the establishment of a relatively functional division of labour.[93] Although chairs in physics were first awarded to theorists before the First World War[94] it was not really until the 1920s that this particular sub-field managed to establish itself as a distinct reputational organization which controlled resources[95] and, not surprisingly, this was not achieved without some conflicts and bitterness, especially in Germany.[96] The overall growth of US physics and the availability of Rockefeller-funded post-doctoral fellowships for theoreticians greatly helped this establishment[97] which merits a little further discussion.

The search for unity and coherence is a major characteristic of modern physics, if not indeed its dominant value.[98] The mathematization of nature has been a crucial constituent of the overall programme of western science for some centuries, especially in what Karpik terms the 'sciences of discovery',[99] and so it is not too surprising that the importance of formal coherence and mathematical unity should lead eventually to the separate and superior installation of theoretical, mathematical physics. However, as many writers have pointed out, the functional division of labour implied by this separation of theoretical work from experimental work is unique to physics[100] and is fairly recent. This suggests that there are some particular qualities about this field – which only became institutionally established and distinct in the nineteenth century[101] – which enabled this differentiation to become established.

In its combination of technical control, problem uniformity and stability, and high mutual dependence, contemporary physics is unique. As already mentioned, the relatively low degree of task uncertainty enables the field to expand considerably through specialization and differentiation without fragmenting into separate and independent organizations. However the high interdependence of groups and problem areas arising from the combination of centralized and concentrated research technologies,[102] prestigious journal space and funding agencies with large numbers of practitioners – results in their co-ordination and integration becoming a key activity for the organization as a whole, and since reputations in this field are highly scientifically and socially prestigious, the continued operation of the organization as a reputational system is of greater concern to its members. Consequently those who make general sense of the diverse and varied contributions from the large number of researchers become influential and prestigious and the generality of important problems and phenomena is also very high. Because of the considerable degree of theoretical and strategic certainty in physics, compared to other scientific fields, this co-ordination and integration activity can be carried out by scientists who are separate from the actual production and assessment of experimental results just as the co-ordination and control of production in mass production industries is a separate activity from the direct supervision of the production process.[103] Thus the high degree of formalization and standardization of technical procedures and theoretical problems and approaches enabled and co-ordination to be achieved 'at a distance' by a separate group of people through the formal communication system. By reorienting the dominant beliefs of physics in such a way as to emphasize the dependence of experimental results and problems upon theoretical models and formalization, the theoreticians managed to achieve considerable autonomy from, and dominance over, their experimental colleagues.[104]

To some extent this sort of field resembles Woodward's account of large batch and mass production industries in which distinct and·substantial sub-units have to co-ordinate their work in order to achieve organizational success while pursuing their own sub-unit interests.[105] This leads to conflict and disputes combined with highly structured and formalized work processes and communication patterns. In the case of modern physics, increased size has led to considerable differentiation of sub-units which are co-ordinated through an elaborate and highly formal communication systems into a hierarchy of prestige and importance. Sub-unit identification is high – as witnesses by the apparent ease of distinguishing sub-disciplines and specialisms[106] – and many physicists seek reputations with these units much as managers pursue their 'careers' within specialized functions. Yet the interdependence of specialism is also and required 'managing' because resources are limited and are collectively controlled by the organization as whole. Restricted opportunities for participating in the most prestigious activities – and thus obtaining high reputations – increases competition over access to facilities and disagreements about the relative merits of competing expansions and extensions of the 'theory-core'. Thus the various sub-fields compete over the significance of their problems and concerns for the discipline as a whole within the dominant theoretical framework as they struggle to establish the legitimacy of their interests.[107] Concentration of resources such as apparatus, technicians, research funds, and highly valued journal space has resulted in the various sub-fields of physics being highly interdependent as they try to demonstrate their importance for the field to claim centrality and scientific status. The formal communication system ensures that results and ideas are transmitted quickly and effectively but, as in mass production, this is not adequate for co-ordination where there are conflicting interests. Specialization enables physicists to produce original work within the dominant framework but each area has to prove its importance for 'real' physics if it is to provide prestigious reputations. Thus the centrality of sub-fields remains an important issue and how the core is separated from peripheral, or 'deviant'[108] areas is of critical concern to physicists. Indeed their policing of intellectual boundaries and fears of 'pollution' from 'applied' issues and goals are as marked as those of economists, especially in European countries, and are a strong contrast to the more relaxed attitudes of chemists.[109]

SUMMARY

The points made in this chapter can be summarized in the following way.

1. Combining the two aspects of task uncertainty with the two aspects of mutual dependence to differentiate types of scientific fields generates sixteen possibilities.
2. However, nine of these sixteen are unstable and unlikely

to become firmly established. The seven remaining types have quite distinct internal structures and patterns of intellectual organization. They can be described as: *(a)* fragmented adhocracies, *(b)* polycentric oligarchies, *(c)* partitioned bureaucracies, *(d)* professional adhocracies, *(e)* polycentric professions, *(f)* technologically integrated bureaucracies, and *(g)* conceptually integrated bureaucracies.

3. The major differences between these seven types can be summarized under two main headings: the configuration of task and problem areas, and co-ordination and control processes. The first includes the degree of specialization and standardization of tasks and skills, the degree of segmentation of problem areas, the degree of formation of district schools of research, and the extent of hierarchical organizations of sub-units around a single set of significance standards. The second includes the impersonality and formality of co-ordination of results and control procedures, the importance of theoretical co-ordination and strategies, and the scope and intensity of conflicts between scientists.

4. Fragmented adhocracies are fluid and weakly bounded systems of work organization and control with little stable internal differentiation and highly personal co-ordination processes. Research is rather divergent and idiosyncratic in these fields and limited in its interconnectnedness. Although the scope of conflicts is potentially high, its intensity is mitigated by the low degree of mutual dependence and need to interrelate task outcomes and research strategies.

5. Polycentric oligarchies organize research around distinct schools which are dominated by reputational leaders controlling critical resources. Co-ordination remains highly personal but greater mutual dependence leads to more theoretical co-ordination of research within each school and increased emphasis upon demonstrating their overall importance to collective goals. Conflicts are therefore quite intense.

6. Partitioned bureaucracies separate analytical work from empirical applications of the dominant theoretical

framework and accord the former much higher prestige. Analytical skills are highly standardized and work is controlled through an elaborate formal communication system but empirical applications are ambiguous in their meanings and significance. Sub-units develop around such applications but remain subservient to the central theoretical core which firmly controls intellectual boundaries and reputations.

7. Professional adhocracies have extensive specialization and standardization of tasks and skills but varied problems and goals which are not ordered into a single hierarchy of significance. Results are co-ordinated through the formal communication system, but overall integration of strategies is limited. Segmentation does occur around particular problems and goals but organizational boundaries are not very firm or strongly policed. The relative abundance of resources, often from varied sources pursuing varied goals, ensures that competition over priorities and significance criteria is not very intense.

8. Polycentric professions combine some segmentation of problem areas and functional dependence between specialists with the formation of separate research schools. High strategic dependence here implies competition over significance criteria and the importance of convincing others of the importance of particular problems and approaches. Theoretical co-ordination of results and strategies is here more critical to reputations.

9. Technologically integrated bureaucracies organize research around separate problems which are co-ordinated by standardized rules and apparatus. Stable sub-fields function as semi-autonomous reputational organizations which share background assumptions, significance criteria, and technologies. A low degree of strategic dependence means that the percise relationships between these sub-units and their relative importance are not critical issues and so a high degree of specialization and segmentation can exist without co-ordination problems arising.

10. Conceptually integrated bureaucracies are highly rule-governed and strongly bounded organizations which

control and direct research through a very formal and standardized reporting system and a hierarchical reputational structure. Relatively stable sub-units form around particular problems and goals but the high degree of strategic dependence restricts their autonomy and increased the degree of competition between them for centrality to the field. This dependence upon the whole field for high reputations heightens the need to integrate and order contributions from different sub-units and so the importance of theoretical co-ordination of strategies.

Notes and references

[1] See, for example, W. Leontief, 'Theoretical Assumptions and Nonobserved Facts', *American Economic Review*, 61 (1971), 1–17; S. Pollard, *The Wasting of the British Economy*, London: Croom Helm, 1982, ch. 7; R. Nelson and S. G. Winter, *An Evolutionary Theory of Economic Change*, Harved University Press, 1982, pp. 6–48.

[2] P. Deane, 'The Scope and Method of Economic Science', *Economic Journal*, 93 (1983), 1–12, at p. 6.

[3] Deane, loc. cit.

[4] As reflected in differences based on the degree of consensus over paradigms; for example, J. B. Lodahl and G. Gordon, 'The Structure of Scientific Fields and the Functioning of University Graduate Departments', *American Sociological Review*, 37 (1972), 57–72.

[5] Although my usage is not quite the same as his, see H. Mintzberg, *The Structuring of Organisation*, Englewood Cliffs, New Jersey: Prentice-Hall, 1979, pp. 432–442.

[6] As in pre-1933 German psychology and British social anthropology after the Second World War. See, for instance, M. Ash, 'Wilhelm Wundt and Oswald Külpe on the Institutional Status of Psychology', in W. G. Bringmann and R. D. Tweney (eds.), *Wundt Studies*, Toronto: Hogrefe, 1980, 396–421 and 'Academic Politics in the History of Science: Experimental Psychology in Germany, 1879–1941', *Central European History*, 14 (1981), 255–86; A. Kuper, *Anthropologists and Anthropology, The British School 1922–1972*, Harmondsworth: Penguin, 1975, ch. 5.

[7] On artificial intelligence see J. Fleck, 'Development and Establishment in Artificial Intelligence', in N. Elias *et al.* (eds.), *Scientific Establishments and Hierarchies*, Sociology of the Sciences Yearbook 6, Dordrecht: Reidel, 1982.

[8] K. Knorr-Cetina, *The Manufacture of Knowledge*, Oxford: Pergamon, 1981, pp. 81–3.

[9] As sketched by G. Geison, 'Scientific Change, Emerging Specialties and Research Schools', *History of Science*, XIX (1981), 20–40.

[10] Much of the debate between the 'Naturalists' and the 'Experimentalists' in biology in the early twenth century was concerned with competing ideals about 'science,. See, for example, G. Allen, 'Naturalists and Experimentalists: the Genotype and the Phenotype' *Studies in the History of Biology*, 3 (1979), 179–209; 'The Rise and Spread of the Classical School of Heredity, 1910–1930; development and influence of the Mendelian chromosome theory' in N. Reingold (ed.), *Science in the American Context; new perspectives*, Washington D. C.: Smithsonian, 1979a: 'Transformation of a Science: T. H. Morgan and the Emergence of a New American Biology', in A. Oleson and J. Voss (eds.), *The Organization of Knowledge in Modern America, 1860–1920*, John Hopkins University Press, 1979b. On the nature of reductionism in the *Drosphila* programme, see N. Roll-Hansen 'Drosphila Genetics: A Reductionist Research Programme', *Journal of the History of Biology*, 11 (1978), 159–210.

[11] R. Collins, *Conflict Sociology*, New York: Academic Press, 1975, p. 513.

[12] As the NAS report states, in chemistry 'every possible measurement that can yield value is required . . . . all areas of chemistry . . . are concerned with structure . . . . the qualitative understanding of chemistry is almost always in terms of structural units', National Academy of Science, Committee for the Survey of Chemistry, *Chemistry: Opportunities and Needs*, Washington D. C.: National Academy of Sciences, 1965, p. 42, See also p. vii of the same report. Shinn has commented on the concern of mineral chemists with 'the most palpable and directly measurable feature of a given group of elemental substances' in T. Shinn, 'Scientific Disciplines and Organizational Specificity: the Social and Cognitive Configuration of Laboratory Activities' in N. Elias *et al.* (eds.), *Scientific Establishments and Hierarchies*, Sociology of the Sciences Yearbook 6, Dordrecht, Reidel, 1982, p. 254.

[13] As in late nineteenth and early twentieth century physics. See, for example, W. O. Hagstrom, *The Scientific Community*, New York: Basic Books, 1965, pp. 247–52; National Academy of Sciences, Physics Survey Committee, *Physics: Survey and Outlook*, Washington D. C.: National Academy of Sciences, 1966, p. 75.

[14] Compare: S. Bonder, 'Changing the Future of Operations Research', *Operations Research*, 27 (1979), 209–24.

[15] As reported by N. Mullins, 'The Distribution of Social and Cultural Properties in Informal Communication Networks among Biological Scientists' *American Sociological Review*, 33 (1968), 786–97.

[16] See, for instance, D. E. Chubin, 'The Conceptualisation of Scientific Specialties', *The Sociological Quarterly*, 17 (1976), 448–76; K. Studer and D. Chubin, *The Cancer Mission*, London: Sage, 1980, ch. 2; R. Whitley, 'Types of Science, organizational strategies and patterns of work in rsearch laboratories in different scientific fields', *Social Science Information*, 17 (1976), 427–47.

[17] T. S. Kuhn, 'Second Thoughts on Paradigms', in F. Suppe (ed.), *The*

Structure of Scientific Theories, University of Illinois 1974; D. Crane, *Invisible Colleges*, Chicago University Press, 1972; N. Storer, *The Social System of Science*, New York: Holt, Rinehart Winston, 1966.

[18] See the references in notes 8 and 16, and also B. Latour and S. Woolgar, *Laboratory Life*, London: Sage, 1979.

[19] See N. Mullins, *Theories and Theory Groups in Contemporary American Sociology*, New York, Harper & Row, 1973; 'Theories and Theory Groups Revisited', in R. Collins (ed.), *Sociological Theory*, London: Jossey-Bass 1983. Compare, though, E. A. Tiryakian, 'The Significance of Schools in the Development of Sociology' and N. Wiley, 'The Rise and Fall of Dominating Theories in American Sociology' both in W. Snizek *et al.* (eds.), *Contemporary Issues in Theory and Research, A Metasociological Perspective*, London: Aldwych Press, 1979.

[20] On radio astronomy, see: D. Edge and M. Mulkay, *Astronomy Transformed*, New York: Wiley, 1976; a contrasting view of the degree and role of controversy in radio astronomy in the early 1950s is provided by B. Martin, 'Radio Astronomy revisited: a reassessment of the role of competition and conflict in the development of radio astronomy', *Sociological Review*, 26 (1978), 27–55. On late nineteenth century physiology, see G. Allen, *Life Science in the Twentieth Century*, Cambridge, University Press, 1978, ch. 4; G. L. Geison, *Michael Foster and the Cambridge School of Physiology*, Princeton University Press, 1978, pp. 331–51.

[21] As reported by H. M. Collins, 'The Seven Sexes: A Study in the Sociology of a Phenomenon', *Sociology*, 9 (1975), 205–24, 'Son of Seven Sexes: the Social Destruction of a Physical Phenomenon', *Social Studies of Science*, 11 (1981), 33–62.

[22] See D. A. Mackenzie, *Statistics in Britain, 1865–1930*, Edinburg University Press, 1981, ch. 6.

[23] See, for example, Mullins, op. cit., 1973, note 19, pp. 184–202; N. Mullins, 'The Development of Specialties in Social Science: the Case of Ethnomethodology', *Science Studies*, 3 (1973), 245–73.

[20] As discussed by Galtung, see J. Galtung, 'Structure, Culture and Intellectual Style: an essay comparing saxonic, teutonic, gallic and nipponic approaches', *Social Science Information*, 20 (1981), 817–56. Compare H. G. Johnson, 'National Styles in Economic Research', *Daedalus*, 102 (1973), 65–74.

[25] For a brief comparison of authority patterns in USA universities in different fields, see J. M. Beyer and T. M. Lodahl, 'A Comparative Study of Patterns of Influence in United States and English Universities', *Administrative Science Quarterly*, 21 (1976), 104–29.

[26] See the papers by Ash, op. cit., 1980 and 1981, note 6.

[27] At least since the 1950s in the USA. See R. D. Whitley, 'The Development of Management Studies as a Fragmented Adhocracy', *Social Science Information*, 23, 1984.

[28] This is not unique to the human sciences. The efflorescence of the bio-medical scientists since the war has produced similar phenomena and in ecology an exponential increase in membership of the Ecological Sciety of America since 1950 has been accompanied by greatly increased heterogeneity and the formation of seven 'invisible college', according to R. P. McIntosh, The Relationship between Succession and the Recovery Process in Ecosystems', in J. Cairns jnr (ed.), *The Recovery Process in Damaged Ecosystems*, Ann Arbor, Michigan: Ann Arbor Science Publishers, 1980.

[29] Thus, theoretical discussions tend to have little contact with empirical work and and appear to operate largely in isolation from it in recent British sociology. See, for instance, P. Abrams, 'The Collapse of British Sociology?' in P. Abrams, *et al.* (eds.), *Practice and Progress: British Sociology, 1950–1980*, London: Allen & Unwin, 1981; R. A. Kent, *A History of British Empirical Sociology*, Aldershot: Gower, 1982, chs. 5 and 6. In United States' sociology 'theory' has become just another specialism. See, for example, S. Levantman, 'The Rationalisation of American Sociology' and R. L. Simpson, 'Expanding and Declining Fields in American Sociology', both in E. A. Tiryakian (ed.), *The Phenomenon of Sociology*, New York: Appleton-Century-Crofts, 1971.

[30] As chronicled by P. Abrams, *The Origins of British Sociology: 1834–1914*, University of Chicago Press, 1968; R. A. Kent, op. cit., chs. 1 and 6.

[31] The importance of undergraduate teaching in British universities, and the relative lack of postgraduate students in the human sciences, are important and underestimated factors in the development of intellectual diversity and lack of cohesion in some fields which underwent a large expansion of posts.

[32] For psychology, see M. Ash, op. cit., 1981, note 6. who criticizes the well-known account of J. Ben-David and R. Collins, 'Social Factors in the Origins of a New Science: the case of Psychology', *American Sociological Review*, 31 (1966), 451–65. On the high growth of junior staff relative to full *ordinarius* positions in the late nineteenth century in medicine and natural science see C. W. McClelland, *State, Society and University in Germany, 1700–1914*, Cambridge University Press, 1981, pp. 258–80; P.Lundgreen, 'The Organization of Science and Technology in France: a German perspective', in R. Fox and G. Weisz (eds.), *The Organisation of Science and Technology in France, 1808–1914*, Cambridge University Press, 1980, pp. 316–20; F. Ringer, 'The German Academic Community', in A. Oleson and J. Voss (eds.), *The Organization of Knowledge in Modern America, 1860–1920*, Johns Hopkins University Press, 1979, pp. 419–20.

[33] As sketched by N. Elias, 'Scientific Establishments' in N. Elias *et al.* (eds.), *Scientific Establishments and Hierarchies*, Sociology of the Sciences Yearbook 6, Dordrecht, Reidel, 1982.

[34] As did Liebig in organic chemistry (*Annalen der Chemie and Pharmacie*), Foster in physiology (*Journal of Physiology*), and Wundt in psychology (*Philosophische Studien*).

[35] Ash, op. cit. 1981, note 6; see also U. Geuter, 'The Uses of History for

the Shaping of a Field; Some Observations on German Psychology' in L. Graham *et al.* (eds.), *Functions and Uses of Disciplinary Histories*, Sociology of the Sciences Yearbook 7, Dordrecht: Reidel, 1983. Psychology was also subordinated to philosophy at Harvard for some time, see B. Kuklick, *The Rise of American Philosophy*, Yale University Press, 1977, pp. 459–63.

[36] As reported by M. Ash, op. cit., 1980, note 6, p. 410. See also G. Böhme, 'Cognitive Norms, Knowledge-Interests and the Constitution of the Scientific Object: a Case Study in the Functioning of Rules for Experimentations', in E. Mendelsohn *et al.* (eds.), *The Social Production of Scientific Knowledge*, Sociology of the Sciences Yearbook 1, Dordrecht: Reidel, 1977.

[37] A. Kuper, op. cit., 1975, note 6, p. 161.

[38] *Idem*, ch. 5.

[39] See the flood of papers and books attacking 'positivism' and similar evils in the social sciences since 1960; also see D. Willer and J. Willer, *Systematic Empiricism*, Englewood Cliffs, New Jersey: Prentice-Hall 1973; Kent, op. cit., 1982, note 29, chs. 5 and 6; Levantman, op. cit., 1971, note 29.

[40] G. W. Stocking, 'The Scientific Reaction Against Cultural Anthropology, 1917–1920', in his *Race, Culture and Evolution*, New York: Free Press, 1968.

[41] As Thompson suggests, all organizations seek to insulate and buffer their core technologies to ensure that 'technical rationality' is optimal. See J. D. Thompson, *Organisations in Action*, New York: McGraw-Hill, 1967, pp. 18–27.

[42] As in economics, compare Deane, op. cit., 1983, note 2; T. W. Hutchison, *Knowledge and Ignorance in Economics*, Oxford: Blackwell, 1977, ch. 4.

[43] See, for example, T. W. Hutchison, op. cit., 1977, ch. 2; J. Hicks, *Causality in Economics*, Oxford: Blackwell, 1979, Preface and ch. 1; Leontief, op. cit., 1971, note 1; H. Katouzian, *Ideology and Method in Economics*, London: Macmillan, 1980, ch. 3; V. J. Taraschio and B. Caldwell, 'Theory Choice in Economics: Philosophy and Practice', *Journal of Economic Issues*, XIII (1979), 983–1006; J. A. Swaney and R. Premus, 'Modern Empiricism and Quantum Leap Theorizing in Economics,' *Journal of Economic Issues*, XVI (1982), 713–30.

[44] See, for example, T. W. Hutchison, *On Revolutions and Progress in Economic Knowledge*, Cambridge University Press, 1978, ch. 11; H. Katouzian, op. cit., 1980, ch. 4; G. Routh, *The Origin of Economic Ideas*, London: Macmillan, 1975, ch. 1; G. Stigler, *The Economist as Preacher*, Oxford, Blackwell, 1982, ch. 17.

[45] See, for instance, M. O. Furner, *Advocacy and Objectivity, a crisis in the professionalisation of American Social Science, 1865–1905*, University of Kentucky Press, 1975, pp. 258–60; H. Katouzian, op. cit., 1980, note 39, ch. 5; R. Nelson and S. Winter, op. cit., 1982, note 1, 6–27.

[46] The domination of undergraduate instructions in Anglo-Saxon economics by a few textbooks which made their authors wealthy does not seem to be paralleled to nearly the same extent by the other human sciences. According to Rosen: 'Virtually every college graduate who has taken a course in economics in the past twenty years has come under the influence of Samuelson's *Economics* and its imitators'; he also notes that this one book had sold over two million copies in the USA alone over twenty years. See S. M. Rosen 'Keynes without Gadflies', in E. K. Hunt and J. G. Schwartz, *A Critique of Economic Theory*, Harmondsworth, Penguin, 1972, p. 401. On the use of rhetoric in Samuelson's textbook see D. N. McCloskey 'The Rhetoric of Economics', *Journal of Economic Literature*, XXI (1983), 481–517.

[47] T. S. Kuhn, *The Essential Tension*, Chicago University Press, 1977, pp. 179–92 and 228–39.

[48] On the 'arithmomorphic' nature of orthodox economics, see N. Georgescu-Roegen, *The Entropy Law and the Economic Process*, Harvard University Press, 1971, ch. 11: P. Jenkin, *Microeconomics and British Government in the 1970s*, Manchester University, unpublished PhD thesis, 1979, ch. 2. On the marginalist 'revolution', see M. Blaug, 'Was there a Marginal Revolution?' and A. W. Coats, 'The Economic and Social Context of the Marginal Revolution of the 1870s', in R. D. Collison Black *et al.* (eds.), *The Marginal Revolution in Economics*, Durham, N. Carolina: Duke University Press, 1973; T. W. Hutchison, *On Revolution and Progress in Economic Knowledge*, Cambridge University Press, 1978, chs. 3 and 4. George Stigler has linked the rise of marginalism to the professionalization of economics in the USA in his 'The adoption of Marginal Utility Theory' in the Collison Black volume and Mary Furner has suggested that marginalism aided the restriction of topics in the move to non-contraversial problems in economics in the 1880s and 1890s and concomitant professional status in her *Advocacy and Objectivity*, University of Kentucky Press, 1975.

[49] Kuhn, op. cit., 1977, note 47, p. 229.

[50] Compare M. Shubik, 'A Curmudgeon's Guide to Microeconomics', *Journal of Economic Literature*, 8 (1970), 405–34.

[51] On the growth of mathematics in economics see H. Katouzian, op. cit., 1980, note 43, ch. 7; see also T. W. Hutchison, op. cit., 1977, note 42, ch. 4; Johnson, op. cit., 1973, note 24.

[52] See, for example, L. A. Boland, *The Foundations of Economic Method*, London: Allen & Unwin, 1982, ch. 5; B. Caldwell, *Beyond Positivism*, London: Allen & Unwin, 1982, chs. 7 and 8; S. Latsis, 'A Research Programme in Economics' in S. Latsis (ed.), *Method and Appraisal in Economics*, Cambridge University Press, 1976; T. W. Hutchison, op. cit., 1977, note 42, ch. 2; F. Machlup, 'Theories of the Firm: Marginal Behavioural and Managerial', *American Economic Review*, 57 (1967), 1–33. New economic approaches to firm behaviour are in N. Kay, *The Evolving Firm*, London: Macmillan, 1982 and Scott Moss, *An Economic Theory of Business Strategy*, Oxford: Martin Robertson, 1981.

[53] P. Jenkin, op. cit., note 47, p. 47; cf. T. W. Hutchison, *The Politics and*

Philosophy of Economics, Oxford: Blackwell, 1981, ch. 9; Kay, op. cit., 1982, ch. 1; B. J. Loasby, *Choice, complexity and ignorance,* Cambridge University Press, 1976, chs. 2, 3, 4., McCloskey, op. cit., 1983, note 46.

[54] Leontief, op. cit., 1971, note 1.

[55] As in the Leontief 'paradox' concerning the Heckscher–Ohlin theory of trade. See, for example, H. Katouzian, op. cit., 1980, note 39, pp. 69–70. For an interesting attempt at demonstrating the 'progressiveness' of theoretical economists who ignored Leontief's result using Lakatos's framework see: N. de Marchi, 'Anomaly and the Development of Economics: the case of the Leontief paradox', in S. Latsis (ed.), *Method and Appraisal in Economics,* Cambridge University Press, 1976. A less favourable application of the Popperian school's ideas to economics is in T. W. Hutchison, op. cit., 1977, note 42, ch. 3.

[56] Hutchison, op. cit., 1978, note 44, p. 319, suggests that contemporary economics is now 'too massive . . . too extensive, sprawling and decentralised . . . to be transformable as an entirety' and 'The busy new suburbs sprawling out in all directions . . . go their own ways largely in intellectual independence of the run-down city centre of 'general theory', but this view ignores the high standardization of analytical skills and domination of theoretical ideals in academic economics. The expansion of jobs and funds has had much more fragmentary impact in US sociology than it did in US economics although there is certainly more discussion of the difficulties of the orthodoxy in the journals today than there was in the 1950s and 1960s. Compare Deane, op. cit., 1983, note 2; Nelson and Winter, op. cit., 1982, note 1.

[57] In the sense of Pantin's distinction between the 'restricted' and 'unrestricted' sciences, see, C. F. A. Pantin, *The Relations between the Sciences,* Cambridge University Press, 1968, ch. 1.

[58] H. Martins, 'The Kuhnian "Revolution" and Its Implications for Sociology', in T. J. Nossiter *et al,* (eds.), *Imagination and Precision in the Social Sciences,* London: Faber & Faber, 1972; C. Lammers, 'Mono-and Polyparadigmatic Developments in Natural and Social Sciences', in R. Whitley (ed.), *Social Processes of Scientific Development,* London: Routledge & Kegan Paul, 1974.

[59] Compare Mullins, op. cit., 1973, note 19, pp. 216–41, with Mullins, op. cit., 1983, note 19.

[60] Knorr-Cetina, op. cit., 1981, note 8, pp. 81–3; B. Latour and S. Woolgar, op. cit., 1979, note 18.

[61] These techiques attempt to reduce the raw material of biological research to standard entities in a similar way to nineteenth-century chemistry; cf. A. Rip, 'The Development of Restrictedness in the Sciences', in N. Elias *et al.* (eds.), *Scientific Establishments and Hierarchies,* Sociology of the Sciences Yearbook 6, Dordrecht: Reidel, 1982. Such attempts result in the scientific labelling of problems and ordering of priorities diverging substantially from medical ones.

[62] The proliferation of diverse names and titles of organizationl units in the biological sciences has become especially marked since the influx of funds for bio-medical research. See, for example, N. C. Mullins, op. cit., 1968, note 15; National Academy of Sciences, Committee on Research in the Life Science, *The Life Sciences*, Washington D. C.: National Academy of Sciences, 1970, pp. 230–9, 282.

[63] On national differences in the development of biochemistry, see R. Kohler, 'The History of Biochemistry: a survey' *Journal of the History of Biology*, 8 (1975), 275–318.

[64] This distinction is taken from the Netherlands and is not echoed in Britain. I am indebted to drs T. Baal for this point.

[65] As become apparent in some interviews with researchers in a cancer research laboratory conducted in the early 1970s. Compare R. Whitley, 'Types of Science, Organizational Strategies and Patterns of Work in Research Laboratories in different Scientific Fields', *Social Science Information*, 17 (1978), 427–47.

[66] Latour and Woolgar, op. cit., 1979, note 18, chs. 2 and 4.

[67] As reported by James Fleck, *The Structure and Development of Artificial Intelligence*, Manchester University, unpublished MSc dissertation, 1978; 'Development and Establishment in Artificial Intelligence', in N. Elias *et al* (eds.), *Scientific Establishments and Hierarchies*, Sociology of the Sciences Yearbook 6, Dordrecht: Reidel, 1982.

[68] As heralded by the Lighthill report, see Sir James Lighthill, 'Artificial Intelligence: a General Survey', in Science Research Council, *Artificial Intelligence: a Paper Symposium*, London: SRC 1973.

[69] At least in the last thirty years or so. See C. S. Fisher, 'Some Social Characteristics of Mathematicians and Their Work', *American Journal of Sociology*, 78 (1973), 1094–118; W. O. Hagstrom, op. cit., 1965, note 11, pp. 226–35; L. L. Hargens, *Patterns of Scientific Research*, Washington D. C.: American Sociological Association, ASA Rose Monograph Series, 1975.

[70] Fisher, op. cit., 1973.

[71] Hagstrom, op. cit., 1965, note 13, pp. 227–35.

[72] See, for instance, G. Allen, op. cit., 1979b; note 10; E. Yoxen, 'Giving Life a New Meaning: The Rise of the Molecular Biology Establishment', in N. Elias *et al.* (eds.), *Scientific Establishments and Hierarchies*, Sociology of the Sciences Yearbook 6, Dordrecht; Reidel, 1982.

[73] As analysed by Yoxen, among others, at any rate. See E. A. Yoxen, 'Life as a Productive Force: Capitalising the Science and Technology of Molecular Biology', R. M. Young and L. Levidow (eds.), *Studies in the Labour Process*, London: CSE Books, 1981; op. cit., 1982. See also R. E. Kohler, 'Warren Weaver and the Rockefeller Foundation Program in Molecular Biology', in N. Reingold (ed.), *The Sciences in the American Context*, Washington D. C.: Smithsonian Institution Press, 1979.

[74] In this case a new conceptualization of the central phenomenon, 'life',

became dominant which was linked to new techniques and competence criteria. Without the prestige of physicists and their instruments, plus the backing of some élite physiologists – at least in the UK – it seems unlikely this change would have occured so easily. Compare E. Yoxen, op. cit., 1981.

[75] Discussed in G. Geison, *Michael Foster and the Cambridge School Physiology*, Princeton University Press, 1978, pp. 331–3.

[76] *Idem*, pp. 176, 330–1.

[77] *Idem*.

[78] As discussed by G. Allen, op. cit., 1979b; note 10; 1979a, note 10. Thus, conflicting groups of experimental geneticists and paleontologists could become united around the work of population geneticists in the 1930s. See W. B. Provine, 'The Role of Mathematical Population Geneticists in the Evolutionary Synthesis of the 1930s and 1940s', *Studies in History of Biology*, 2 (1978), 167–92.

[79] Geison, op. cit., 1981, note 9.

[80] As discussed, for instance, in Mintzberg, op. cit., 1979, note 5.

[81] G. Bachelard, *La Formation de l' Esprit Scientifique*, Paris: J. Vrin, 8th ed., 1972, p. 61.

[82] Compare W. O. Hagstrom, op. cit., 1965, note 13, p. 246.

[83] Shinn, op. cit., 1982; note 12.

[84] Georgescu-Roegen, op. cit., note 48, pp. 114–23.

[85] Compare D. A. Bantz, 'The Structure of Discovery: Evolution of Structural Accounts of Chemical Bonding', in T. Nickles (ed.), *Scientific Discovery, Case Studies*, Dordrecht: Reidel, 1980.

[86] R. G. A. Dolby, 'Debates over the Theory of Solution', *Historical Studies in the Physical Sciences*, 7 (1976), 297–404.

[87] See M. Crosland, 'Chemistry and the Chemical Revolution', in G. S. Rousseau and Roy Porter (eds.), *The Ferment of Knowledge*, Cambridge University Press, 1980. In any case chemistry traditionally was linked to medicine and pharmacy, see K. Hufbauer, 'Social Support for Chemistry in Germany during the 18th Century: How and Why did it Change?' *Historical Studies in the Physical Sciences*, 3 (1971), 205–31.

[88] Discussed by, among others, B. Gustin, *The Emergence of the German Chemical Profession, 1790–1867*, University of Chicago, unpublished PhD thesis, 1975, chs. 5, 6, and 7; D. Kevles, 'The Physics, Mathematics and Chemistry Communities: A Comparative Analysis', in A. Oleson and J. Voss (eds.), *The Organization of Knowledge in Modern America, 1860–1920*, John Hopkins University Press, 1979; H. W. Paul, 'Apollo courts the Vulcans: the applied science institutes in nineteenth-century French science faculties', in R. Fox and G. Weisz (eds.), *The Organization of Science and Technology in France, 1808–1914*, Cambridge University Press, 1980.

[89] Compare W. O. Hagstrom, op. cit., 1965, note 13, p. 160.

[90] Compare the distinction between 'intensive' and 'extensive' physics in

the NAS Physics Survey Committee, *Physics in Perspective*, Washington D. C.: National Academy of Sciences, 1972, pp. 581–5.

[91] Compare N. Georgescu-Roegen, op. cit., 1971, note 48, chs. 4 and 5; C. F. A. Pantin, op. cit., 1968, note 57, ch. 1.

[92] NAS Physics Survey Committee, op. cit., 1972, p. 364.

[93] See J. Gaston: *Originality and Competition in Science*, Chicago University Press, 1973, pp. 58–9; Hagstrom, op. cit., 1965, note 13, ch. 5

[94] P. Forman, J. Heilbron, and S. Weart, *Personnel, Funding and Productivity in Physics circa 1900*, Historical Studies in the Physical Sciences, 5 (1975), pp. 31–9.

[95] See, for example, S. Weart, 'The Physics Business in America, 1919–1940', in N. Reingold (ed.), *The Sciences in the American Context*, Washington D. C.: Smithsonian Institution Press, 1979, p. 300.

[96] Where it was linked to anti-semitism, see Forman *et al*, op. cit., 1975, note 94, pp. 31–2. See also P. Forman, 'Alfred Landé and the Anomalous Zeeman Effect, 1919–1921' *Historical Studies in the Physical Sciences*, 2 (1970), 153–261; E. Crawford, 'Definitions of scientific discovery as reflected in Nobel prize decisions, 1901–1915', paper presented to the 1981 annual meeting of the Société française de Sociologie, Paris.

[97] As documented by Weart, op. cit., 1979, note 95.

[98] Compare J. Heilbron, 'Fin-de-Siècle Physics', paper presented to the 1981 Nobel Symposium on Science, Technology and Society in the time of Alfred Nobel, Stockholm. See also P. Galison, 'Re-Reading the Past from the End of Physics', in L. Graham *et al*. (eds.), *Functions and Uses of Disciplinary Histories*, Sociology of the Sciences Yearbook VII, Dordrecht: Reidel, 1983; NAS Physics Survey Committee, op. cit., 1972, note 90, pp. 57–75.

[99] L. Karpik, 'Organization, Institutions and History' in L. Karpik (ed.), *Organization and Environment*, London: Sage, 1979, pp. 18–23.

[100] Hagstrom, op. cit., 1965, note 13, pp. 247–52.

[101] See, for example, R. Silliman, 'Fresnel and the Emergence of Physics as a Discipline', *Historical Studies in the Physical Sciences*, 5 (1974), 137–62; S. F. Cannon, *Science in Culture*, New York: Science History publications, 1978, ch. 4.

[102] Especially in nuclear and elementary particle physics but also in other areas where elaborate facilities were required for research on electromagnetism and radiation. The average construction costs of physics institutes rose by 80 per cent between the 1890s and 1905–1914 (Forman *et al*, op. cit., 1975, note 94, p. 102). Despite Rutherford's pride about constructing apparatus with primitive and cheap facilities, physicists increasingly used commercially produced instruments at the turn of the century (*idem*, pp. 85–6).

[103] Compare J. Woodward, *Industrial Organization*, Oxford University Press, 1980, pp. 136–45; A. L. Stinchcombe, 'Bureaucratic and Craft

Administration of Production', *Administrative Science Quarterly*, 4 (1959), 168–87.

[104] As reflected in Gaston's study of British high energy physicists; Gaston, op. cit., 1973, note 93, pp. 30–1.

[105] Woodward, op. cit., 1980, chs. 8 and 11.

[106] As in the NAS report, op. cit., 1972, note 90, ch. 4.

[107] Hagstrom, op. cit., 1965, note 13, pp. 168–70.

[108] As in Harvey's account of research on hidden variables in quantum mechanics. See B. Harvey, 'The Effects of Social Context on the Process of Scientific Investigation: Experimental Tests of Quantum Mechanics', in K. Knorr *et al.* (eds.), *The Social Process of Scientific Investigation*. Sociology of the Sciences Yearbook 4, Dordrecht: Reidel, 1980.

[109] Compare, for instance, the different approaches taken by the two NAS survey committees to the utilitarian justification of research in chemistry and physics. See also Crawford's account of conflicts within physics over Nobel Prize allocations in contrast with the chemists' acceptance of a rotation principle – E. Crawford, op. cit., 1981, note 96. The growth of academic and industrial physics in the 1920s led to an increasing separation between 'pure' and 'applied' physics which became reflected in the publication policy of the *Physical Review* and the establishment of 'applied' journals. Increasingly the most prestigious journal became dominated by articles in 'pure' nuclear physics and other areas were relegated to separate journals. At the same time, the top twenty graduate schools increased their domination of US physics as it expanded. See S. Weart, op. cit., 1979, note 95. In contrast chemistry does not seem to have drawn distinctions so sharply nor to have organized itself into a hierarchy of 'fundamentality'. Compare Hagstrom, op. cit., 1965, note 13; Hargens, op. cit., 1975, note 69.

6

THE CONTEXTS OF SCIENTIFIC FIELDS

INTRODUCTION

JUST as particular combinations of degrees of mutual dependence and task uncertainty are associated with different internal structure of scientific fields, so too they occur in different contexts. Changes in these contexts can similarly be expected to be associated with changes in the way research is organized and controlled and so in the particular type of scientific field most representative of a given subject area. For example, the growth in the amount of funds provided for research, and in the variety of institutions providing them, in the biological sciences over the past forty years or so have increased the diversity of audiences for task outcomes and the goals pursued by researchers in many fields so that the degree of strategic dependence between scientists has declined considerably in some of the traditional disciplines, to the extent that they have ceased to function as the dominant reputational organizational in many bio-medical fields.[1] In contrast, the development of 'big science' around very expensive and complex apparatus in other fields has encouraged the establishment of more formal collaboration and co-ordination structures and restricted the degree of theoretical diversity which might otherwise have developed.[2]

In Chapter 3 I suggested that there are three major sets of contextual factors which affect the degree of mutual dependence between scientists and in Chapter 4 these were related to differences in the degree of task uncertainty. In this chapter I discuss these factors, their interrelationships and some sources of variation in them in greater detail before analysing their connections with the seven major types of scientific field identified in Chapter 5.

CONTEXTUAL FACTORS AFFECTING THE STRUCTURES OF
SCIENTIFIC FIELDS

The three sets of contextual factors mentioned in Chapters 3 and 4 were: *(a)* the degree of reputational autonomy from competing intellectual organizations and the wider social structure in setting standards, *(b)* the degree of concentration of control over access to the means of knowledge production and validation, and *(c)* the structure of reputational audiences. These in turn were subdivided into distinct components in the following way. Reputational autonomy covers three separate areas of control: performance standards, significance standards, and descriptive terms and concepts. Concentration of control over key resources has two major sub-dimensions based on the external and internal relations of employment units–horizontal and vertical concentration. Audience structure also has two components–the variety of audiences available to members of a field seeking positive reputations, and the extent to which such audiences are ranked in terma of their prestige and inportance–their reputational equivalence. These seven dimensions will now be considered, together with some interconnections between them and some reasons for differences in them.

(a) Reputational autonomy. The first dimension, the degree of reputational control over competence and' performance standards in scientific research, is critical to the ability of a scientific field to control work processes and labour markets and so the nature of competent contributions to knowledge in that field. Some minimal level of autonomy in setting such standards is therefore essential for any reputational organization to become established as a separate scientific field. However, sciences manifestly do differ in the extent to which their leaders are able to decide how well a piece of work was done without reference to employers' criteria or the views of other groups, and these differences are related to the degree of co-ordination and comparison of task outcomes in scientific fields.

Three distinct levels of reputational autonomy over setting performance standards can be identified. First, a low level of reputational control exists in a field when there are no clear work methods which are unique to that field and researchers trained in other sciences are able to make recognized contributions to intellectual goals without substantially modifying their procedures. British sociology and management studies seem to be examples of this situation. A medium level of control exists when there is a distinct set of 'craft' procedures entrenched in training programmes and communication channels which differentiate competent research from 'amateur' contributions so that adequate performance is assessed in terms of the correct use of disciplinary methods for appropriate problems. An example of this level of control might be the use of particular participant-observer techniques in British social anthropology in the 1950s and 1960s. A high level of control over performance standards exists when the reputational élite sets competence criteria in isolation from other groups so that no contribution to intellectual goals can be made without acquiring research skills in their training establishments. What constitutes a competent piece of research is here largely controlled by the professional group which is able to reject and ignore work produced with other skills and procedures. Anglo-Saxon economics manifests this level of reputational control over analytical skills.

The second dimension, control over significance standards, deals with the assessment of problems and research strategies rather than with research skills and abilities. It is the overall importance and primacy of problems which are controlled by these standards so that they order contributions according to their centrality to collective objectives. Reputational control of these standards is less critical to organizational identity and survival than are performance standards, but total domination of them by external groups implies an inability to order research questions and strategies around distinct programmes and priorities so that intellectual coherence and systematic direction of knowledge development is impossible. Total subservience to external influences on these standards means that the field has no cental issues or problems and so no clear intellectual identity.

A low degree of reputational autonomy in setting significance standards exists in a field when groups in other areas, including lay ones, are able to influence substantially collective views about what the most important problems are and the extent to which contributions deal with them successfully. The direction of research goals in these fields is subject to the values and purposes of employers, funding agencies, and élites from other disciplines to the extent that scientists redirect their strategies and evaluation criteria if these groups alter their priorities as seems to be the case in much bio-medical research.

Control over significance standards is greater where the fields' leaders have some means of insisting in its own intellectual goals and ranking achievements in terms of their contributions to these distinct objectives. The importance of problems and topics in fields with a medium degree of reputational autonomy is assessed mostly by specialist colleagues, but their standards are significantly affected by those current in other fields and by general societal priorities. An example of this is the role of philosophers in the early development of psychology.[3] A high degree of autonomy in controlling significance standards exists when reputational leaders are able to exclude other groups from influencing evaluations of contributions except in so far as they provide funds and facilities. What counts as a major piece of research is determined by the values and goals of the fields which also have a major effect on how resources are allocated so that the level of external control in setting intellectual objectives is low. Much of modern physics exemplifies this level of reputational control of significance standards.

These standards are linked to the third aspect of reputational autonomy: control over the field's characterization of its domain, its problems, and descriptive language. The ability of an organization to determine its own boundaries and what is, or is not, an appropriate problem for its members to work on is clearly crucial to its identity. Where scientific fields construct their domains in a largely residual manner, i.e. out of whatever other fields have omitted, such as sociology in the USA at the end of the nineteenth century[4], they are highly constrained by existing boundaries and so subject to other groups' definitions and purposes. Equally, where the central phenomenon of a field is largely defined by powerful non-scientific groups, as in management studies, its autonomy is very limited and subject to invasion in the name of 'relevance'.[5] A less extreme situation is where the major phenomenon are redefined and boundaries redrawn as a result of members of other fields developing new research programmes around new descriptions of central objects, such as 'life' in the biological sciences.[6] It is important to note here that this aspect of a field's environment does not only refer to its boundaries and key problems but also covers the concepts and vocabulary used to describe phenomena and task outcomes. This is especially crucial in an organization so dependent upon an international communication system to function at all. Where descriptive terms are very close to common sense and subject to multiple interpretations, as in the human sciences, it is difficult to reduce ambiguity sufficiently to be able to rely entirely on the formal communication system for controlling research through reputations and to integrate results around common theoretical goals.

A low degree of reputational control over domain descriptions, problem formulations, and descriptive terms in a field exists when everyday language is used to describe phenomena, and technical definitions of concepts are accorded no superior status over commonsense usages such as in many of the human sciences. Also, what constitutes a disciplinary problem or topic is subject to other groups' pressures so that the ability of scientists in the field to claim a monopoly of particular issues and concerns is distinctly limited. Rather, what issues are 'in' or 'out' of the field is difficult to discern and the object of extensive and frequent negotiations within and without the field. A medium degree of autonomy exists when technical terms and concepts are entrenched in training programmes, and are regarded as superior to everyday terminology, but there is still a considerable overlap between them and common-sense usages so that confusion can arise over cognitive objects and the implications of research results. Similarly, although the domain of the field is more distinctly bounded and policed here, it is subject to lay pressures and legitimation criteria. What is, or is not, an economic problem is largely determined by the analytical framework of neoclassical economics but external influences do affect the range of issues tackled and policy demands lead to extensions of that framework to new areas which were not previously considered appropriate.[7] A high degree of reputational control exists when technical terms dominate the communication system and lay concepts are excluded. The domain of a field is here also free from lay formulations of problems but is affected by the boundaries and strategies of more powerful scientific groups as well as being subject to financial pressures. The critical point, though, is that what constitutes a disciplinary problem or topic, and how it should be conceived, are controlled by reputational leaders rather than by outsiders as in 'pure' mathematics where applications are accorded low prestige.[8]

These three areas of reputational autonomy are interconnected in that where performance standards are subject to other groups' criteria it is unlikely that significance standards and descriptive terms will also be unaffected by such criteria. In the employee-dominated sciences it is the control of skills and how performance is assessed which is critical for control over problem importance and descriptive concepts for reputational groups seeking autonomy and independence. [9] Once they can exert some influence on how knowledge is to be defined and assesssed in the field, and on how the appropriate knowledge production and validation skills are to be delineated and inculcated, they can begin to control labour markets and develop a distinctive intellectual framework. In claiming resources for training programmes and research facilities, new disciplines have to demonstrate their ability to reproduce distinctive skills which are essential for producing the distinctive sort of knowledge characteristic of the emerging field.[10] Without some such unique skills it becomes very difficult to control research procedures and evaluation criteria so that a field is liable to be taken over by other groups—both scientific and lay—and fragmented into separate sub-units controlled by more established groups. This seems to be the situation of much modern British sociology.[11]

While the lack of control over performance standards limits the ability of reputational groups to control significance standards and domain boundaries, the existence of some degree of reputational autonomy in this area does not, of course, automatically imply the strong reputational control of all aspects of research. As with other professional groups, the extent to which any control over skills can be combined with control over work goals differs considerably in the sciences and this has become especially noticeable with the expansion of the number and variety of employers of scientific expertise for knowledge production. Within the public sciences many fields share control over work goals with a variety of groups and agencies so that the traditional notion of the academic discipline combining training programmes, skill certification, and a monopoly of jobs and access to facilities is applicable to only part of modern science. Autonomy from external influences over significance standards and descriptive concepts and boundaries, thus varies between fields even when they have considerable control over skill production and definition.

These latter variations in control are interconnected in that a high degree of reputational control over significance standards is improbable without a fair degree of control over domain boundaries and descriptive terms. Where, for instance, cognitive objects and descriptions are closely related to everyday discourse so that it is difficult to distinguish technical usages from common-sense ones, reputational groups will find it difficult to establish their own criteria for assessing the significance of contributions because alternative descriptions are available and these can be linked to alternative priorities as in many of the human sciences. Additionally, it will be unclear how research results contribute to overall goals if they are formulated in diffuse and relatively ambiguous terms. A low degree of external influence upon significance standards implies, then, some separation of technical terms from everyday usages and some ability to control admissible contributions to intellectual debate and its boundaries.

The case of Anglo-Saxon economics is an interesting instance of this point. Here, there are clear boundaries between what is, and what is not, an economic problem and what are admissible ways of proceeding to deal with such problems. Skills are highly standardized and controlled by the reputational élite so that performance standards are highly autonomous. Yet, the topics which economics purports to deal with are important everyday issues and the terminology of economic analysis is close to ordinary language even if technical definitions of these terms are quite distinct. Profit, for example, has a technical meaning in micro-economics which is not the same as usages in conventional accountancy – which themselves vary. In their internal discussions and rankings of intellectual significance of contributions economists use their technical concepts and own standards of evaluation, yet they also make policy pronouncements and seek to intervene in everyday debates and discussions where terms are not technically defined and they do not control usages.[12] Indeed, it seems unlikely that economics would receive so much financial support if it was not thought that its subjects and problems were strongly connected with everyday phenomena. Thus, within the professional fraternity, performance and significiance standards, and technical terminology, are fairly strongly controlled by the reputational élite but there is also a considerable overlap with common-sense terms and concerns which legitimates public support and sometimes affects standards.[13] As with the problem of demonstrating effective technical control over phenomena, this difficulty is resolved to some extent in economics by separating analytical economics from fields of application and elevating the former to a position of dominance and insulating it from external influences to a very high degree. Thus economics is a hybrid science in which divergent features are combined so that the core exhibits different characteristics to the peripheral subfields.

Such interconnections between these three areas of reputational autonomy reduce the number of likely empirical combinations of different degrees of autonomy in them considerably. However, the degree of reputational control over standards operates more as a constraint rather than as a determining factor. While a high degree of reputational control over performance standards is a prerequisite for the establishment of scientific fields with high strategic dependence and low task uncertainty, it does not follow that these latter characteristics will always develop in these circumstances. In mathematics, for example, criteria for assessing competence and significance are largely controlled by academic mathematicians, and their descriptive language is highly esoteric and separate from everyday discourse, yet this field is quite different in its dominant pattern of work organization and control from modern physics and chemistry. Also, strong influences from diverse funding agencies and other professional groups can prevent high reputational control of significance standards and theoretical integration but their absence does not imply a strong ordering of sub-units in terms of overall importance and theoretical unity. Equally, the same degree of external influence on all three areas of reputational control can be associated with different internal structures as in the cases of modern physics and chemistry. Both of these sciences have a high degree of autonomy over all three areas and yet they differ in their degree of strategic dependence and importance of theoretical co-ordination. This is largely due to variations in the degree of concentration of control over access to means of intellectual production and the structure of their audiences.

(b) Concentration of control over access to the means of intellectual production and distribution. The division of concentration of control over access to the means of intellectual production and distribution into two distinct aspects – horizontal and vertical – parallels that discussed by Mintzberg in his analysis of how organizations either decentralize decisions away from line management to staff experts (horizontal) and/or from top management to supervisors and operators (vertical).[14] However, here I am concerned with the extent to which control over jobs, facilities, funds, and journal space is dominated by a small group from a small number of employment units and research sites (horizontal concentration) and to which it is unequally shared between employees within those units (vertical concentration).

The first aspect refers to the extent to which scientists in each work-place are able to pursue their own interpretation of organizational goals with distinct research strategies and programmes through local control of resources. In highly horizontally concentrated fields control over resources is exercised through a small number of central agencies allocating grants, funds for posts and equipment and resources for training recruits, so that a small élite group is able to set research goals and order reputations for contributions to collective goals. An obvious example of such a field is high energy physics where scientists have to convince a relatively small group of colleagues of the importance and significance of their research to gain access to the requisite facilities. In contrast, research in many human and biological sciences can be conducted with more limited resources which are readily available in most employment organizations and so these exhibit a lower degree of horizontal concentration. Generally, the more major resources are concentrated in a single source such as one Research Council, or jobs and research facilities are concentrated in a few élite universities, or space in the most prestigious journal is clearly much more highly valued than space in other journals, the more a field is highly horizontally concentrated. A medium degree of horizontal concentration occurs when some important resources are centrally controlled and differently allocated to research sites but each employment unit is able to exercise some control over goals and strategies. Much research in the modern natural sciences in Britain reflects this situation since funds often are available from only one agency, and are allocated preferentially to only a relatively small number of élite groups associated with dominant training centres, yet journals are not controlled by a central élite and each group pursues different priorities and approaches.[15]

The second aspect is similar to the more familiar notion of centralization of authority in which decisions and rewards are controlled by a small group at the top of some authority hierarchy and operatives have little or no influence on them. In science this refers to the extent to which the research director or professor controls who does what research in what way and when. Not only does this vary across national employment systems, between for example the autocratic powers of the nineteenth-century Prussian *Ordinarius* and the more democratic structure of many North American university departments,[16] but it also differs between scientific fields. Shinn, for instance, has reported considerable variations between mineral chemistry, solid state physics, and computerized vector analysis in the extent to which technicians and senior and junior scientists were involved in planning research and interpreting results.[17] Some fields, such as parts of chemistry, permit a more centralized and bureaucratic style of organizing research in employment units while others, such as branches of applied mathematics, seem to require greater involvement by all members of the group and cannot be controlled by a single authority position through highly formalized procedures.

In general, vertical concentration is high when the leader of an employment unit controls appointments and promotions, allocates local research facilities and support, controls grant applications and the submission of reports and articles for publication. A medium degree of vertical concentration implies local administrative control over some of these resources – usually appointments – but not all, so that scientists are able to pursue different research strategies from the organizational director without being directly penalized. Fellows of Oxbridge colleges in fields where some professorial control of grant applications and publication submissions exists, seem to be in this situation.[18] A low degree of vertical concentration implies relatively little local administrative control of research strategies, facilities, or procedures. Here, researchers are able to pursue their own indications without too much concern for their departmental heads, perhaps because they acquire permanent positions relatively easily and do not need extensive research facilities. Obviously some local control of performance assessment remains, but this does not substantially restrict the sort of research undertaken or how it is done.

Different degrees of these two aspects of centralization can be combined so that strong professorial control over research, for instance, can occur with a lower degree of horizontal concentration. Mid-twentieth-century British anthropology was largely conducted in six university departments, for example, which were firmly controlled by the professors, but major resources were fairly equally distributed between them so each could develop its own distinctive approach to the subject.[19] A similar characterization could be applied to the German university system although the various ministries of education played a much more direct role in appointments and budgets than does the British state and the University of Berlin was often the dominant employer in many areas.[20] In general, vertical concentration seems to be greater in the natural sciences, perhaps because of the greater technical control over phenomena and division of labour,[21] but it can be associated with a medium degree of horizontal concentration as in the biological sciences – or with highly centralized control of resources as in parts of astronomy and physics. Where resources are widely available from a wide variety of agencies pursuing a variety of objectives, as in the bio-medical sciences in the USA, it seems unlikely that organizational leaders will be able to dominate research strategies and procedures to a high degree, and so vertical concentration will be limited.

Both types of concentration are clearly dependent upon national variations in the organization of higher education and employment structures. The British science system, for example, is more centralized, both horizontally and vertically, than is the United States' one.[22] These variations are particularly important in fields where skills and symbol structures are not standardized and technical task uncertainty is relatively high because task outcomes cannot be systematically compared and co-ordinated across research sites and national cultures through the formal communication system and personal connections are critical for making common sense of results. In such fields, national structures of organizational careers and resource allocation mechanisms have major effects on the organization of the reputational system and so on types of intellectual output. Galtung's identification of different intellectual styles in the social sciences,[23] for example, are clearly related to the way jobs, facilities, and other resources are organized in different countries. His suggestion that the Teutonic and Gallic styles of discourse are more homogeneous and less tolerant of diversity than the Saxonic can be partly explained by the greater degree of vertical concentration of scientific research in France and Germany.[24] An additional institutional feature which is relevant here is the nature of the jobs available in different national systems. In the German university system the full professor had to cover a wide range of topics over the whole discipline – and, of course, a substantial part of his income was derived from lecture fees which encouraged him to teach the most general and basic courses[25] – so that a broad competence over many specialized areas was essential to obtain chairs and general theoretical integration of specialized research highly valued. In the Anglo-Saxon system in contrast, tenure became available at an earlier stage so practitioners were not so dependent upon the professor,[26] although this did not mean they were totally independent of him, and full professors have become more numerous and not expected to cover entire disciplines. Specialization was thus more encouraged and the intensity of competition thereby reduced.

In scientific fields with lower technical task uncertainty and more standardized skills and symbol systems, on the other hand, these sorts of national differences have rather less impact upon the control of research and types of intellectual style. Local characteristics and exigencies are less crucial in establishing research strategies and evaluating task outcomes as result are less ambiguous and there is a greater degree of common background knowledge and training. Thus it becomes more feasible to characterize the field as a whole without distinguishing between national variants and styles. The structure of the international reputational organization in these fields plays a more important role in directing research and co-ordinating results relative to national employment structures.[27]

These more 'internationalized' sciences do, though, reflect the dominant characteristics of the most important national system in the field. Just as multinational firms tend to centralize control in the host country and disseminate its managerial culture, so too there are hierarchical relations between national scientific systems in the sciences and one style often dominates – especially in the more centralized ones. While the German science system was probably the most important one for many fields in the nineteenth and early twentieth centuries, the dominant one today is the United States' and its national characteristics increasingly permeate how research is done and how it is assessed in sciences with an international publication system and standardized skills. This domination is considerably aided by the transmission of such standardized skills to peripheral centres by extensive migration to the central country for training. The importance of Germany as a scientific training centre, especially in chemistry, for North Americans and other foreigners in the nineteenth century is well known,[28] while the domination of Notrh American graduate schools since the 1940s has become obvious. Increasingly, research skills and strategies have become determined by the pattern of education and employment characteristic of the United States in many fields as Europeans and others have flocked there for their training and certification as competent practitioners. Just as publishers have become more dependent upon the United States' market for sales and, therefore, editorial decisions, so too it is one's reputation in the American academic system which determines scientific standing and hence national reputations in many sciences. In so far as that system tends to promote highly specialized, empirically specific research through relatively low degrees of horizontal and vertical concentration, then this becomes the dominant style of scientific work.

Just as we can distinguish between fields in terms of their horizontal concentration within a country, so too we can differentiate their degree of international horizontal concentration. Where one country dominates research priorities and facilities to the extent that its scientists are able to set significance and performance criteria for the whole international system of research, horizontal concentration is obviously quite high. In such fields, scientists are primarily oriented to the reputational system of one country and are primarily governed by its priorities and significance criteria. They seek publication in its journals rather than in their own and compete for positions in organizations there, if only temporary ones. Where international horizontal concentration is lower, on the other hand, national variations in intellectual priorities and styles of work are more important and noticeable and scientists are more oriented to their national colleagues than to the international reputational system. Local élites and organizational leaders are more crucial in these fields for individuals' reputations and access to facilities so that personal standing among national *patrons* is the major focus of research strategies. Here, if the national system is highly concentrated, both horizontally and vertically, scientists will be highly dependent upon colleagues in that system but obviously largely autonomous from international opinion. This autonomy is perhaps most visible in French intellectual life,[29] especially in the human sciences, but can also be observed in some English and Dutch groups where local political and literary repute is more highly valued than international scholarly reputations.

A related source of variation in concentration of control over reputational organizations in the employee-dominated sciences is the general availability of jobs and the conditions under which they are allocated. The more these are tied to reputational criteria, the stronger the influence of the national and international élites, but, equally, the easier jobs are to obtain the weaker that influence becomes as theoretical deviants manage to find posts. This is especially so when researchers do not need to obtain outside funding from a single source for their work and when journal space is not in very short supply. The actual number of scientists and research sites competing for reputations in a field seems to be critical here, as does the extent to which research centres are strongly controlled by a single leader. Where the number of contributors is fairly small, as in most of the human sciences before the expansions of the 1960s, and each work-place is firmly under the intellectual and administrative control of the professor, a strong professorial establishment emerges, as in Germany, which sets standards and co-ordinates research. Although no single employer dominates the field and there are a number of variations between the centres as each struggles to establish its vision of the discipline as the primary one, research is strongly mutually oriented and scientists focus on colleagues rather than alternative audiences. If, on the other hand, jobs and facilities are not so directly controlled by the professor, as in the nineteenth-century Oxford and Cambridge universities – and still so to some extent – this strong mutual orientation declines as scientists specialize around separate topics and form distinct reputational units which are not so highly co-ordinated by a relatively small professorial élite. Concentration and hence mutual dependence are here lower.

(c) Audience structure. The third set of contextual features of scientific fields to be considered here concerns the audiences for research results which lead to reputaions. The first component of this factor simply refers to the degree of variety of distinct groups, to which scientists may address results for reputational purposes and the diversity of their goals and evaluation criteria. In the bio-medical sciences, for example, many research projects produce results which can be published in a wide variety of journals oriented to diverse groups of scientists, administrators, and doctors. In physics, on the other hand, it is usually fairly clear which journal is appropriate for any one result. Audiences in this latter field are much more limited and specific to topics and problems.[30]

Diversity of audiences is high where scientists are able to obtain reputations from lay groups as well as from groups of colleagues with similar skills, such as in English literary studies, but is also considerable when different groups of scientists pursuing distinct intellectual goals are available as potential sources of legitimate reputations in the field, such as recent social psychology. Scientific fields with a large number of different and diverse audiences are, then, only weakly bounded and share control of work goals – and perhaps procedures – with other groups so that theoretical integration is unlikely to be high. Audience variety is low, on the other hand, when scientists have a limited number of specialist groups to address for reputational purposes and these share common intellectual goals, as in much of modern chemistry where research is highly specialized and tied to particular groups of specialist colleagues.

The second component, audience equivalence, refers to the extent to which different audiences are roughly equal in respect of the reputations they control. Scientists may have a number of different groups to address with the outcomes of a particular project, but some may be much more influential and important than others so that audiences have a low degree of reputational equivalence as in economics. Thus publication media may be ranked in terms of the size and importance of the audience they reach and influence – which is why general journals are often more prestigious than specialist ones – and so competition for access to the most general medium is strong. In fields where audiences are more equivalent there is not a clear and commonly accepted hierarchy of importance of audiences or journals so that competition for the attention of a particular group or space in a particular medium is relatively low. Such a situation implies a low degree of mutal dependence and little need for practitioners to co-ordinate their research strategies or results.

In principle, scientific fields can have varying degrees of audience variety associatd with varying degrees of audience equivalence. A small number of audiences, such as groups defined by increasing specialization of interests within a stable theoretical and technical environment, may be quite strongly ranked in terms of their importance and the prestige of the reputations they control. Alternatively, diverse and numerous audiences may also be associated with some ordering of their importance, as in economics where lay groups may be interested in the results of forecasting models but scientific status is clearly controlled by professional colleagues.[31] Additionally, highly specialized and restricted audiences may not always be ranked in terms of their relative importance to disciplinary goals and so may be roughly equivalent as far as reputations are concerned. Specialist groups in twentieth-century chemistry seem to reflect this situation. Finally, diverse and plural audiences are quite likely to occur with little ranking of their reputations as in many of the human sciences where the 'educated public' is a legitimate source of renown and senior posts are awarded to practitioners who have reputations in other fields[32] or whose books are more influential in extra-mural circles than among disciplinary colleagues. Audience diversity in such fields is combined with high equivalence to reduce reputational control of research strategies and procedures so that mutual dependence is low and task uncertainty high.

Scientific fields with greater control over performance standards than fragmented adhocracies and some skill standardization can more easily exclude lay audiences as being incompetent to judge contributions. Here, audience diversity is therefore lower but the number of potential audiences may still be considerable if different research sites follow separate strategies and horizontal concentration is limited. Furthermore, audiences from other fields which are more prestigious in the scientific system as a whole may be available and may be influential in ordering significance standards. The importance of philosophical criteria and ideas in the development of German psychology is an instance of this[33] as is the continuing influence of philosophical debates and points of view in many human sciences.

Turning to the sources of audience diversity and equivalence, or of changes in them, we can see that an obvious factor is the plurality and diversity of employment opportunities and access to research facilities and funds. Where researchers are able to obtain resources from a variety of agencies and organizations, they are less likely to be restricted to a single group for reputations or to accept a strong hierarchy of audiences and their goals in setting significance standards. Another factor is the relative prestige of the field in the overall ordering of the sciences. If it is relatively low then scientists may well be tempted to seek reputations from related audiences in more prestigious areas and impose their standards upon colleagues. Here, the variety of audiences increases, but they may also become more highly ordered if such attempts are successful through substantial additional resources becoming available and being monopolized by these groups.[34] Where a field is highly prestigious on the other hand, audiences are unlikely to be plural and diverse as the disciplinary élite can exclude divergent groups and influences and maintain its own priorities and standards by offering more valued reputations. The more concentrated is access to the necessary resources for intellectual production, the more likely specialist groups in such prestigious fields are hierarchically ordered and so audience equivalence is low.

The diversity of intellectual audiences is also dependent upon the degree to which leaders of the scientific field are able to control the definition of its phenomena and standards of competent research practices. Where commonsense and everyday terminology is used to structure problems and characterize task outcomes, non-scientific groups are obviously able to influence interpretations of results and offer alternative assessments. In principle, then, they can function as possible audiences for researchers. If, though, the reputational organizaion is able to control jobs and other necessary resources through its control system then audiences will be sufficiently unequal to reduce external influences. In many of the human sciences, for instance, task outcomes are written in everyday language but professional groups are often able to insist on their standards of competence and significance so that lay assessments carry relatively little weight in appointments and promotion decisions. Audiences here are plural and diverse but their evaluations are not equivalent. This downgrading of lay opinions and goals is easier when the number of employers is small and so personal connections can be used to ensure collegiate control over appointments and/or when skills are sufficiently standardized and esoteric that special training can be insisted upon as a requirement for practitioners to participate in professional debates, and this training is controlled by the reputational élite of the field. Such standardization itself, of course, encourages the development of a more esoteric communication system which effectively reduces lay ability to assess intellectual products as in the 'mathematical revolution' in economics in the 1950s and 1960s.[35]

This exclusion or devaluation of lay opinion and criteria is easier to accomplish when the major topics and concerns of the field are not critical to the interests of powerful non-scientific groups; or, if they are, then scientists can directly demonstrate their ability to be useful to those interests as in the case of nuclear weapons and energy. The professionalization of United States' economics in the last decades of the nineteenth century has been linked by some authors to its eschewal of controversial issues as an instance of this process.[36] It also required substantial reputational control over resources so that employers and funding agencies had to influence research goals and standards through the reputational system rather than being able to do so directly. This implies that jobs and training programmes are not controlled by lay groups pursuing non-intellectual goals. However, simple autonomy from direct non-scientific control of goals and performance assessment does not in itself ensure skill standardization or exclusion of lay values, as the case of

Anglo-Saxon academic sociology indicates. Reductions in audience diversity and equivalence seem to require, in addition, increased competition for reputations arising, for instance, from increased numbers of practitioners, and/or greater inequality of control over valued resources – such as jobs at prestigious organizations – and so increased concentration of collegiate control, together with increased reputational control over training programmes and the certification of competences, if the degree of skill standardization is to increase. Where a common craft skill is not essential for producing competent contributions to collective intellectual goals, audience diversity will remain high. Laboratory techniques and practices in psychology, for instance, enabled a distinct competence to be institutionalized which excluded lay contributions, but the social survey techniques which reigned over US sociology for a time no longer serve the same function, and so it is unclear what is the distinctive competence of sociologists that requires extensive advanced education and which enables them to deal with distinctive problems.

THE CONTEXTUAL STRUCTURES OF SEVEN MAJOR TYPES OF
SCIENTIFIC FIELD

These seven components of the contexts of scientific fields together account for many of the major differences between the seven major types of science identified in the previous chapter. Each type of scientific field develops and becomes established in different contextual circumstances as summarized in Table 6.1. The particular features which are characteristic of each field will now be briefly discussed, together with some likely consequences of changes in them. As before, the examples mentioned are illustrative and not intended as detailed analyses of particular areas of work.

(a) Fragmented adhocracies. This type of scientific field develops and continues as a distinct organization when reputational control over standards and concepts is limited and shared with other scientific and lay groups. The nature of the research skills required to produce competent contributions to knowledge in these fields is subject to a variety of influences, and work procedures are not specific to a single field, nor are

TABLE 6.1
The Contexts of Seven Major Types of Scientific Fields

Types of Scientific Field	Contextual Factors						
	Reputational Autonomy over setting:			Concentration of control over access to critical resources		Audience structure	
	(a) Performance standards	(b) Significance standards	(c) Problem formulations and descriptive terms	(a) Horizontal	(b) Vertical	(a) Audience variety	(b) Audience equivalence
(a) Fragmented adhocracy	Low	Low	Low	Low	Low	High	High
(b) Polycentric oligarchy	Medium	Medium	Low	Medium	High	High	Medium
(d) Partitioned bureaucracy	High	High in core Low in periphery	High in core Low in periphery	High in core Medium in periphery	Medium	Medium	Low
(e) Professional adhocracy	High	Low	Medium	Low	Low	High	Medium
(f) Polycentric profession	High	Medium	High	Medium	High	Medium	Medium
(g) Technologically integrated bureaucracy	High	High	High	Low	High	Low	Medium
(h) Conceptually integrated bureaucracy	High	High	High	High	High	Low	Low

they controlled by a particular set of reputational leaders. In many of the human sciences, for instance, distinct craft skills are supplemented or dominated by general norms of 'good scholarship' which are diffuse and broadly applicable to many areas. What counts as competence is open to many interpretations and this diffuseness enables considerable local variations to develop. It also allows élite groups to apply standards particularistically in ways which are difficult to challenge. Judgements of intellectual competence and performance in these fields are relatively open to lay influences and criteria drawn from more prestigious fields, as when eminent laymen review books in general literary journals and mathematical sophistication becomes regarded as a major property of research competence. This plurality of influences upon performance standards restricts the development of standardized research procedures and skills, and also limits the need and ability to co-ordinate and compare systematically task outcomes throughout the field.

The low degree of reputational control of performance standards in fragmented adhocracies implies an equally low degree of specialist control over significance standards. The formulation and ranking of research problems in these fields are subject to a wide variety of criteria and goals from employers, funding agencies, general cultural fashions and preferences, and from other scientific fields. Where élite groups from other spheres of activity impinge upon significance criteria, scientists are encouraged to pursue a wide range of interests and intellectual goals without having to co-ordinate research strategies or justify them to particular reputational groups. They deal with a variety of problems in a variety of ways and seek reputations for diffuse contributions to broadly conceived goals without demonstrating their specific connections to specialist colleagues' strategies. With limited collective control over what counts as an important problem or how problems should be formulated, research becomes rather disconnected and fragmented as scientists pursue distinct interests and topics in idiosyncratic ways. The openness of fragmented adhocracies to diverse influences and goals also leads to considerable fluidity and instability of research goals and ideals as fashions change and reputations fluctuate. The rise and decline of approaches in Anglo-Saxon sociology and literary studies, and eclectic empiricism of much research in management studies exemplify these processes.[37]

This openness to general cultural élites and goals is facilitated and echoed by the use of everyday discourse to conduct and report research in these fields. Cognitive objects, tools, and task outcomes are characterized in ordinary language so that results and goals are accessible to the educated laity and thus to their judgements. The development of standardized and esoteric terminology is threatened by alternative usages legitimized by élite groups from other areas of intellectual work. Characterizing sociologists' language as 'jargon ridden', for instance, is a way of denying the legitimacy of esoteric analysis of social phenomena and asserting the subservience of sociological work to general cultural norms and standards. This exoteric language system in fragmented adhocracies encourages diffuse contributions which are open to varied interpretations and so difficult to compare and contrast systematically across research groups and historical periods. Co-ordination of task outcomes around common goals and problems is therefore limited. This is not to argue that the establishment of a technical language in itself will increase intellectual cohesion – if that is desired – but simply to point out that relying upon ordinary language restricts the extent to which a reputational organization can maintain intellectual boundaries and co-ordinate research procedures and results. As long as research goals and products are accessible to the lay public, and diffuse, general cultural standards are legitimate criteria for assessing contributions, it is improbable that research in fragmented adhocracies will become highly co-ordinated or systematically interrelated.

The limited control over standards and concepts exercised by any one reputational group in these fields is reflected in, and the outcome of, the relatively low degree of concentration of control over access to critical resources exercised by reputational leaders. Typically, working materials are either cheap and easy to appropriate personally, as in the case of personal libraries, or else are a collective facility controlled by larger organizations than individual departments such as university libraries, so that particular reputational élites cannot dominate access to them. The development of laboratory research in the nineteenth century, which enabled institute leaders to monopolize access to essential facilities, heralded a major change in the degree of concentration of control, and thus the degree of mutual dependence between scientists. Additionally, access to jobs and major communication media is not monopolized by a single élite in fragmented adhocracies but rather is available to specialist groups pursuing a variety of goals with different approachers. These groups are usually able to influence the allocation of permanent posts without having to co-ordinate their goals or dominant views with other groups, and reputations in one specialism are not greatly superior to those in others.

The degree of vertical concentration is also relatively low in fragmented adhocracies. Research is largely an individual affair with little centralization of direction or evaluation in employment units, and administrative leaders tend to delegate much control over appointments and promotions to specialist reputational groups in these fields such as the sociology of A or B. Thus specialization and differentiation of problems and topic areas along a variety of dimensions is encouraged and there is little need for much internal co-ordination or integration of research strategies. Typically, permanent employment status, or tenure, is granted relatively early in scientists' careers and this restricted degree of vertical centralization in fragmented adhocracies is echoed by a limited reliance upon national funding agencies for producing competent contributions and limited dependence upon highly prestigious journals controlled by a single reputational group. Also, much publication is in book form in these fields which further lowers the degree of concentration of control since publishers' goals have some influence over publication decisions.

The low extent of co-ordination of research results and strategies in these fields is also an outcome of the varied audience available to scientists for reputations and their relative equivalence. Since the boundaries of fragmented adhocracies are fairly fluid and permeable to outside goals and standards, scientists can seek positive reputations for work oriented to a variety of purpose and are not dependent upon a single reputational élite or ranking of significance standards. So, in management studies audiences often comprise a mixture of university employees and 'management intellectuals' from consultancy firms and from large bureaucracies and they fluctuate in prestige.[38] Equally, in British sociology researchers sometimes obtain high reputations for books addressed to the general educated public and/or students rather than for specialist articles focused upon a small group of professional colleagues.[39] A reliance on traditional, broadly based, cultural élites for reputations is, of course a major standpoint in the long running dispute between different approaches in English literary studies which conflict over performance standards, significance standards, and descriptive terms. The legitimacy of such appeals to lay audiences reduces the need to connect task outcomes with those from specialist colleagues and encourages intellectual diversity.

(b) Polycentric oligarchies.   As can be seen from Table 6.1, these sorts of scientific fields became established when reputational leaders can exert greater control over standards and restrict the range of audiences available for legitimate reputations. The high degree of strategic dependence in polycentric oligarchies implies that reputational autonomy from lay groups is considerably greater than in fragmented adhocracies, and specialist reputations count for more than do broader reputations among cultural élites. However, the lack of standardized procedures and raw materials, which is linked to the use of everyday descriptions and ordinary language, restricts the degree of co-ordination of results across research schools and of research strategies.

The major reason for this higher degree of strategic dependence in this type of scientific field is, I suggest, the near monopolization of knowledge production and certification by academics in highly centralized employment units. Polycentric oligarchies develop when a relatively small group of universities or similar institutions dominate both the production and certification of research skills, and the employment of knowledge producers, and these are hierarchically structured under the control of an intellectual leader. The degree of horizontal concentration of control is greater than in the case just considered because a few élite departments dominate the inculcation of essential skills and access to the major communication media, as in the case of U.S. sociology of the 1950s,[40] and the degree of vertical concentration is high because of the near dictatorial powers of the institute heads, as in the traditional nineteenth century German research institute. Reputations in these fields depend upon a relatively small number of people who personally and locally control most of the critical resources for making competent contributions to collective intellectual goals. Even though they compete for domination of the reputational organization, and pursue different strategies, goals, and approaches, together they constitute the intellectual and administrative establishment which controls the direction and evaluation of research. Without certification from one of these major departments and the personal patronage of one leader it is difficult for scientists to make major contributions and gain access to material rewards through reputations. For example, the leaders of the major British anthropology departments in the 1950s and 1960s controlled the allocation of field-work grants for students which were essential for gaining professional expertise and the necessary access to raw materials to make competent contributions.[41]

The importance of concentrating control over graduate education and publications media for increasing the degree of mutual dependence can be clearly seen in the history of North American sociology and anthropology.[42] Chicago, Columbia, and Harvard dominated the production of PhDs in the United States up to the 1960s in sociology and also dominated the discipline although Chicago lost its early publications monopoly when the new *American Sociological Review* was established in 1935.[43] Similarly, Mullins considers the Broom and Selznick textbook as being crucial to the remodelling and standardization of graduate teaching in sociology along structural functionalist lines in the 1950s.[44] The ability of Parsonian sociology, and its Mertonian variant at Columbia, to generate problems for students to work on and so gain reputations is seen as a critical factor in eventual domination of US sociology by many observers. Equally, in anthropology Boas at Columbia was able to dominate graduate training and so control the new departments being established in the 1920s. Again control over journal space was a critical factor in intellectual and political battles in that field during and just after the First World War.[45] In both fields research tended to be dominated by a relatively small number of people located in a relatively small number of centres before the expansion of posts in the 1960s; the system was small enough to be controlled largely through personal contacts and networks based on teacher – student relationships.

Because the high degree of co-ordination and control of research in polycentric oligarchies is highly personal and often dependent upon administrative control of critical resources, it is vulnerable to major changes in the amount and distribution of such resources. In particular, if jobs and materials become much more widely available, and more evenly distributed between individuals and employment units, the ability of a small group of leaders to dominate knowledge production and validation is severely reduced. Thus the expansion of university teaching posts in many industrialized countries in the 1960s and 1970s weakened the hierarchical authority structure of many employment units and the dominant position of élite training schools and so the degree of concentration of control over access to critical resources.

For example, the growth of graduate education and jobs for many social scientists in the USA in the past two decades enabled new centres of skill production and certification to become established,[46] and deviant theoretical groups such as the ethnomethodologists in sociology to obtain and control access to some jobs.[47] It also created a larger market for academic publications which enabled new media to become established that were not controlled by existing reputational leaders. Thus the amount and variety of journal space increased and concentration of control over access to major communication media was reduced. The need to follow established disciplinary leaders to obtain reputations for competent and significant work declined as specialist groups pursuing their own goals and priorities were able to establish novel reputational organizations which exercised some influence over resource allocation standards. Furthermore, the lack of standardized work procedures meant that the enlarged knowledge production system could not be co-ordinated and controlled by a national élite as personal connections and influence became ineffective co-ordination mechanisms. The degree of strategic dependence among scientists declined as a result and the dominant pattern of organization became more similar to fragmented adhocracies.

The present decline in ease of access to critical resources, such as jobs and research funds, in many western countries might increase the degree of concentration of control, and thus the degree of mutual dependence. However, it is by no means certain that the cut-backs in higher education are resulting in greater centralization of control in fragmented adhocracies, nor that reputational organizations in general are becoming more integrated and co-ordinated. Given the existing high degree of pluralism in many human sciences and their ready access to a variety of audiences, it seems unlikely that researchers will respond to a general reduction in the availability of resources by seeking reputations from a single dominant group. Rather, increased competition for jobs seems likely to encourage specialization and topic differentiation without necessarily leading to greater theoretical co-ordination and integration of results across employment units.

(d) Partitioned bureaucracies. These fields develop when reputational groups have quite a high degree of autonomy over performance standards but control over significance standards is shared with powerful external groups in peripheral areas. Equally, problem formulations and descriptive terms are quite esoteric and standardized in the analytical core, yet they overlap with everyday concepts and usages in 'applied' areas. Thus reputational control over work processes and goals is high in the central sub-field but declines as more empirical and specific issues are considered. It therefore varies between sub-units.

This variation also occurs with audience structure. Given the range of areas of 'application' of the dominant framework, there are quite a large number of distinct audiences for research in partitioned bureaucracies outside the analytical core. However, these audiences usually have much less intellectual prestige than does the professional élite which controls theoretical work and its standards. So while audiences for peripheral problems are quite numerous and roughly equivalent, the reputational system as a whole is rigidly hierarchical and audiences are definitely ranked according to a single dimension of intellectual prestige.

Similar variations occur with the degree of concentration of control over access to critical resources. In peripheral areas, funds and other resources are available from a variety of agencies which are not always controlled by the dominant establishment. Thus, many economists can obtain access to raw materials through state statistical bureaux which publish much of the needed data, without being highly dependent upon a particular group of professional colleagues, although personal relationships with the senior employees of such bureaux may often be an advantage. Equally, many researchers in economics can publish their work in 'applied' journals without a great deal of difficulty provided they adhere to the dominant intellectual approach. Horizontal concentration of control is therefore not very high in many peripheral areas.

This is not true, though, of the central core. Because reputations are highest here and lead to the 'best' jobs and honours, competition is intense. Access to core journals is correspondingly highly valued and, since analytical skills are quite highly standardized, is largely controlled by the central establishment through the formal communication system. Deviant or controversial work are rarely published in partitioned bureaucracies where the formal rules determine competence and significance in the core sub-field. The high degree of standardization and formalization of procedures and standards in the central sub-field enables control over access to prestigious journal space to be concentrated in the hands of a relatively small élite group employed at a relatively small number of élite departments.[48]

This high degree of horizontal concentration of control is both reflected in, and increased by, a similar concentration of skill training and certification establishments. In the 1950s, for example, six university departments in the United States, Berkeley, Chicago, Columbia, Harvard, MIT, and Wisconsin, produced nearly one-half of the PhDs in economics and two-thirds of these obtained academic jobs and published articles. Many of these papers were in the central sub-fields and most of the present reputational leadership of present US economics came from those schools.[49] So while competent contributions to intellectual goals can be made by graduates from less prestigious schools, they are usually numerically and qualitatively less important to the reputational system than those from the dominant departments. The likelihood of making highly valued contributions to the most prestigious sub-field is considerably greater, then, for scientists who were trained at, or are members of, the major graduate schools, which tend, in economics, to recruit their own products.[50]

In contrast, the degree of vertical concentration in partitioned bureaucracies is not especially high. The high degree of analytical skill and problem standardization in these fields means that local variations in intellectual priorities and approaches are limited by the dominant reputational organization and the ability of local hierarchies to direct research strategies through controlling access to key resources is constrained by the greater importance of national and international reputations. The power of local 'barons', although not negligible, is restricted by the highly formalized and rule-governed reputational system which determines admissible problems and strategies. If local leaders were able to dominate research strategies and priorities so that scientists followed divergent paths and directions between employment units, they would not be able to publish in the central journals or likely to receive high reputations and so would not influence the field as a whole. If, on the other hand, they worked within the dominant orthodoxy, their subordinates could obtain high reputations and move elsewhere, thus effectively reducing the importance of local, vertical concentration of control over access to critical resources. In general, then, the high degree of standardization of analytical research skills, admissible problems, and ways of dealing with them, and the relatively high prestige of reputations in the core sub-field, mean that partitioned bureaucracies are unlikely to become established where there is a high degree of vertical concentration of control coupled with a low degree of horizontal concentration.

(e) Professional adhocracies. In the remaining four types of scientific field, performance and competence standards are largely determined by reputational leaders who control the inculcation and certification of knowledge production skills in particular sciences. However, other contextual factors vary considerably between them, as Table 6.1. indicates. In professional adhocracies scientists have only limited control over significance standards and work goals, and are open to a variety of influences from other reputational groups and lay agencies. Polycentric professions, on the other hand, have greater autonomy in setting intellectual priorities and are more able to exclude lay terms and concepts. Their audience structure also tends to be less diverse and more ordered in importance and prestige. Both technologically and conceptually integrated bureaucracies are quite highly autonomous in setting standards and controlling languages but differ in their degree of concentration of control over access to critical resources and equivalence of audiences. These differences are related to the relative scarcity of essential resources and their distribution among subfields.

Because performance standards and skills are largely controlled by scientists in professional adhocracies, lay terms and reputational criteria have less impact upon research processes and the evaluation of task outcomes than in fragmented adhocracies. Thus, cognitive objects and research results are more esoteric and specific in bio-medical research than in most sociology. However, the importance of particular problems and evaluation of research goals are subject to external criteria and standards in these fields so that, for instance, changes in funding patterns and agency goals affect intellectual priorities and strategies.[51] The role of non-scientific staff, such as patent lawyers, in setting research priorities in many biological areas has been emphasized by recent studies of research laboratories[52] and demonstrates the plasticity of research strategies in professional adhocracies. Furthermore, the variety of audiences in these fields is quite high and scientists are able to publish in a wide range of journals. Although medical journals and audiences are regarded as being less prestigious and important than 'scientific' ones in neuroendocrinology, they are important for public relations and fund-raising purposes so that writing for them is a legitimate and useful activity.[53] Some audiences, then, are more significant than others, but scientists in these fields are able to seek reputations from a variety of groups oriented to different goals and thus to pursue a variety of intellectual strategies. Additionally, within the more 'scientific' set of journals and audiences, there are a number of possible groups which could be addressed and reputations sought from for any one piece of research and these are not clearly ranked into a stable prestige order according to a single set of significance standards.[54]

This variety of audiences and influences upon significance criteria is linked to the relatively low degree of concentration of control over access to resources in professional adhocracies. Obtaining journal space, research funds, jobs, and facilities is not very difficult for those suitably equipped with certified skills in comparison with polycentric professions. Similarly, although scientists are subject to an administrative hierarchy which allocates apparatus, technicians, and similar resources within employment organizations, their ability to approach funding agencies directly and appeal to the wider reputational system for resources and recognition lessens the influence of local leaders and results in a relatively decentralized process of selecting and formulating research strategies. The diversity and number of funding agencies, employment organizations, and audiences in professional adhocracies mean that researchers have considerable autonomy from any single reputational élite as well as from local leaders so that the levels of both horizontal and vertical concentration of control are relatively low.

This can be seen in the recent expansion of bio-medical research in the USA and its increasing domination by a variety of federal funding agencies, especially the several National Institutes of Health. The capacity of any single academically based élite group to control and direct research strategies in a particular field has become greatly reduced by the large growth in funds, jobs, and facilities which is only indirectly influenced by traditional academic élites. While this pattern of funding agency intervention in setting intellectual priorities and styles of work was established by the Rockefeller foundation in the 1930s, it has since become much more massive and has led to entirely new areas of work being developed by funding agencies.[55] Control over reputations in many biological fields is now shared between academics, funding agency officials, medical élites, and directors of research laboratories who form varied and fluid coalitions dominating what Knorr-Cetina calls 'trans-epistemic arenas of research' rather than scientific communities.[56] The significance of research results in such fields is determined by continuous negotiations between members of various institutions oriented to divergent goals and often changes quite rapidly. Within a shared set of competence standards and general style of research, scientists are relatively autonomous from any one group of specialist colleagues in selecting research strategies so that they do not need to demonstrate the importance and significance of their work for the general collective goals of a particular biological discipline. Co-ordination of research strategies around central objectives in distinct intellectual fields is therefore limited in many US 'life sciences'. Because of different national traditions and structures, this seems to be rather less true of European biomedical research where resources are a little more centrally controlled.[57]

(f) Polycentric professions. Polycentric professions become established in similar contexts to polycentric oligarchies except that they have greater reputational control over competence and performance standards. Their increased control of skills and research procedures is linked to their autonomy from lay terminology and concerns and enables greater co-ordination of results and strategies around common theoretical objectives. Although reputational control of significance standards is not as high as in more bureaucratic fields, it is greater than in sciences where results and goals are expressed in everyday terms and concepts overlap with lay usages. A considerable degree of autonomy and independence from lay goals and standards is therefore necessary for these sorts of fields to become established.

The high degree of mutual dependence in polycentric professions arises from the considerable concentration of control over access to critical resources. Only those people who have the necessary standardized skills are eligible for jobs and access to research facilities and only those who contribute to the collective intellectual goals of individual research schools obtain high reputations and continued access to scarce resources. Essentially, the necessary facilities and opportunities for making competent and highly valued contributions to scientific goals are scarce in relation to the number of qualified practitioners seeking positive reputations in these fields, and access to them is largely controlled by a relatively small set of research leaders pursuing distinct priorities and developing separate approaches. Thus scientists have to persuade their colleagues, especially their leaders, of the competence, significance, and centrality of their contributions and problems if they are to continue to participate in the production of knowledge in their field. Horizontal concentration of control is considerable, then, but not so high as to permit one shool to dominate the entire field. For strategic pluralism and uncertainty to remain substantial, each theoretical approach has to be able to control some resources through its own reputational sub-system. Vertical concentration is greater since each research school co-ordinates and controls strategies and goals around its own programmes, and scientists are constrained to work with in a school rather than being able to pursue a variety of strategies as in professional adhocracies.

Some of the differences between national traditions in mathematics can be illuminated by analysing variations in their degree of concentration and thus their closeness to professional adhocracies on the one hand, or to polycentric professions on the other. In very general terms, mathematicians require few resources for their work compared to other twentieth-century scientific fields – primarily time, i.e. jobs, and access to communication media. In this respect the field is similar to many human sciences. However, skills are highly standardized and significance standards largely controlled by reputational élites who restrict audiences and seek to maintain the high social and intellectual prestige of the field, partly through control of education curricula. Thus reputational autonomy is high and so too is the degree of functional dependence between mathematicians in particular specialisms.[58] However, the degree of strategic dependence, and so too the extent and importance of theoretical co-ordination and integration of specialisms, differs between

North American and continental European mathematics.[59]

In the United States jobs have not been very difficult to obtain, especially since the late 1950s, and employment units are relatively democratically structured so that researchers equipped with certified skills are not highly dependent upon a small number of resources controllers for access to critical resources. As long as they use standard procedures to work on problems in a recognized specialism, mathematicians can obtain positive reputations without needing to demonstrate the overall centrality and significance of their problems to the entire field and thus co-ordinate their work with colleagues in other specialisms. In contrast, Continental employment structures tend to be more vertically concentrated in their allocation of jobs and research time and a relatively small number of professors and other *grands patrons* have tended to dominate national employment opportunities and communication networks.[60] Both vertical and horizontal concentration of control over key resources has been higher than in the United States and so mathematicians have had to convince research leaders of the merits of their contributions to general collective goals, and show how they fitted into the dominant school's approach, if they were to obtain jobs, promotion, and access to journals. The degree of strategic dependence seems, therefore, to be greater in European mathematics which resembles a polycentric profession more than a professional adhocracy. This higher degree of concentration might also be connected to the different prestige and status of applied mathematics and statistics in North America compared to Europe, since it is easier to maintain boundaries and 'purity' when resources are controlled by a relatively small group of administrative and reputational leaders.

Finally, polycentric professions can only become established when the variety of external audiences is limited and, in particular, appeals to lay groups do not lead to positive scientific reputations. If there are a large number of audience oriented to different goals which are available to researchers seeking positive reputations, obviously the degree of dependence upon any single group is not very high and scientists do not need to co-ordinate their research very much with specialist colleagues. However, the existence of distinct schools in these fields implies some variation in intellectual goals and priorities between groups and audiences so that alternatives are available. Equally, while reputations in some specialisms and approaches are probably more prestigious than those in others, so that equivalence of audiences is not total, no single group or approach so dominates significance standards in polycentric professions that it can determine unilaterally the worth of contributions throughout the field. Instead, each specialism and intellectual approach competes for control of the reputational system without any one being able to monopolize it.

(g) Technologically integrated bureaucracies and (h) Conceptually integrated bureaucracies.   These two types of scientific field have many contextual features in common, as can be seen from Table 6.1, so will be discussed together. They share a high degree of reputational autonomy over performance and significance standards as well as over problem formulations descriptive terms. They both also control the definition of cognitive objects and communicate task outcomes through a very esoteric and formalized symbol system. Research is governed by elaborate and general rules which are regarded as invariant between research sites. Thus work carried out across nation states and cultures can be co-ordinated and integrated through the formal communication system by scientists who are personally ignorant of the particular local conditions of intellectual production. Audiences for research results are limited by the esoteric reporting system and the exclusion of scientists from other fields as legitimate sources of reputations. Since the theoretical system is relatively stable and highly elaborated, problems and strategies are formulated in fairly standard ways so that there is little variation in them between employers and research groups and divergent approaches to intellectual goals are unlikely to become established. The variety of goals and approaches available to scientists seeking positive reputations is therefore low.

The major differences in the contexts of these two types of field arise from variations in the concentration of control over access to critical resources and the extent to which some audiences and specialisms are regarded as more prestigious than others. In conceptually integrated bureaucracies resources are scarce relative to the number of qualified scientists seeking positive reputations and are controlled by a relatively small group of élite researchers who collectively determine significance standards and the importance of particular problem areas for the field as a whole. To gain access to critical resources and high reputations, scientists have to demonstrate not only their technical competence and usefulness to specialist goals but also the general importance of their work for the entire discipline and its intellectual goals as interpreted by the élite. Thus specialist groups compete over the centrality and importance of their problems and priorities. Because access to resources is restricted in these fields, the relative significance of each sub-field becomes a matter of considerable concern and precisely how their contributions are to be integrated around broad, general problems important to determine. Environmental pressure thus increases internal co-ordination problems and competition. This competitition, it should be noted, is not over basic intellectual approaches or formulations, since strategic task uncertainty is relatively low, but rather focuses upon the relative contribution which each sub-field makes to the overall goals of the organization and thus upon the relative prestige of reputations in them. In conceptually integrated bureaucracies, then, high levels of horizontal concentration of control over access to key resources lead to a high degree of strategic dependence between scientists and problem areas and so a strong concern with the co-ordination and integration of task outcomes from separate specialisms.

In technologically integrated bureaucracies, on the other hand, resources are more plentiful and more equally distributed among research sites and employment units. Access to jobs, apparatus, technicians, and other important facilities for producing institutionally recognized knowledge is not so concentrated in the hands of a distinct élite but is more generally available to holders of certified skills in each specialist area. While the degree of vertical concentration of control within each employment unit is quite considerable because of the reatively routinized nature of much research and the advantages of hierarchical control and co-ordination of differentiated tasks and skills,[61] horizontal concentration is relatively low and scientists in different area do not need to complete intensively over the relative merits of their problems and concerns. Co-ordination of results and task outcomes occurs through the common background assumptions, theories and apparatus, which are standardized across employment units and training programmes, and does not require a distinct group of intellectual managers to determine presisely the significance of each sub-field. Positive reputations within each specialist group are sufficient to gain access to resources without scientists having to convince the entire field of the centrality of their problems and results so that they do not need to co-ordinate and integrate their work systematically with that of colleagues in other specialist sub-fields.

Many of these differences between technologically and conceptually integrated bureaucracies are observable in the development of modern chemistry and physics. The early involvement of German industry in chemical research, both in firms and educational institutions, provided the basis for an enlarged academic profession in this field in the late nineteenth century and more widely available resources than in physics.[62] As Forman puts it: 'the extraordinary close connection between industry and the academic chemists had long been the envy of the other sciences.[63] Also, the expansion of the number of *privat-dozenten* relative to the number of full chairs in the natural sciences in Germany in the last quarter of the nineteenth century, which intensified competition and concentration of control, seems not to have been so great in chemistry as in other fields.[64] In the 1920s both chemistry and physics expanded rapidly in industry and the universities but chemistry grew at a faster rate and succeeded in attracting more funds from industry.[65] Furthermore, while physics became dominated by atomic physicists and theoreticians who exercised considerable control over post-doctoral fellowships and research funds,[66] chemistry enlarged the range of specialisms and topics covered without apparently concentrating control of critical resources in any single sub-field.[67] Horizontal concentration in physics has, then, been much greater than in chemistry.

The high degree of horizontal concentration of control over acces to critical resources in physics has grown even more since the Second World War with the development of nuclear and elementary particle physics as the dominant sub-field, and corresponding increasing reliance upon expensive and complex apparatus for making competent contributions to these areas.[68] The growth of 'big science' increases concentration of control by centralizing research facilities on an international scale and forcing researchers to justify their strategies and concerns to an élite group in order to gain access to these essential research tools. It is important to emphasize here that this concentration of control is not just a matter of apparatus but also incorporates access to research funds, fellowships, travel grants, and prestigious journal space. The large sums required for much research in physics and astronomy – especially in the most prestigious areas – necessitate reliance upon a small number, often only a single one in European countries, of agencies such as the US Atomic Energy Commission which depend on a small group of élite advisors in allocating resources. In effect this relatively small group now controls the direction of much experimental work in physics since without state funds for apparatus and/or access to state controlled facilities, scientists simply cannot make competent contributions and so gain positive reputations.[69] Furthermore, it is work in the most 'intensive' and expensive sub-fields which dominates the most prestigious journals and leads to the highest reputations in physics so that the overall level of dependence throughout the field upon the élite groups is very high indeed, and competition for access to scare resources correspondingly intense. While many scientific fields allocate space in their more general – and hence often most prestigious – journals to a number of specialist areas, physics has tended to concentrate space and other resources in one dominant area which is the centre of intellectual competition and control.[70] Thus the development of national post-doctoral research fellowships for the 'best' natural scientists, and provision of research posts in élite US universities in the 1920s, enabled quantum theorists to become a dominant group in physics there, and atomic physics to become the central sub-field as it was in Germany.[71]

This high degree of concentration of control has enabled physics to expand greatly without much reducing the degree of mutual dependence between scientists, as has happened in many biological sciences in the past forty years or so. By controlling and restricting problem choice and formulation through a tightly integrated and unified theoretical system which ordered topics in terms of their general importance and excluded or devalued alternative, deviant formulations,[72] the dominant group in physics has been able to redefine problems and priorities from funding groups seeking different objectives into respectable, 'real' physics problems, or else devaluing them as relatively insignificant and shallow.[73] Because of its past success in dealing with military and industrial problems, physics has been able to exercise much more control over the sorts of problems, and how they were formulated and understood, which it was paid to solve than seems to have been the case in the biological sciences. Thus, rather than non-scientific, e.g. medical, goals fragmenting research programmes, and leading to highly diverse problems and strategies, in physics they are co-ordinated and interconnected through the central theoretical structure and its elaborate, hierarchical communication system. The overall scientific and societal prestige of physics is such that this central control is unlikely to be seriously weakened by other scientists developing new descriptions of problems and priorities with the help of outside sponsors as happened in biology. Furthermore, the institutionalization and high prestige of the co-ordination role in a distinct occupational group, theoreticians, means that the dominant values of generality and unification are now built into the control system of modern physics to such an extent that decentralization and diversity of research goals and strategies are unlikely to develop to any significant extent. Rather, the proliferation of sub-fields and problems has increased the demand for theoretical co-ordination and integration so that central control may have actually increased rather than declined as more groups compete for centrality and prestige.

SUMMARY

The points made in this chapter can be summarized in the following way.

1. Different types of scientific field develop and become established in different contexts. Changes in these context

affect the organization and co-ordination of research strategies and problems.

2. The contexts of scientific fields can be analysed in terms of three sets of dimensions: (*a*) the degree of reputational autonomy in setting and controlling performance standards, significance standards, and linguistic structures; (*b*) the degree of concentration of control over access to critical resources, and (*c*) the structure of audiences for reputations.

3. Particular combinations of these factors constrain the development of different types of sciences broadly as follows. The lower the degree of reputational autonomy, the less concentrated is control over access to resources and the more diverse and equivalent are audiences for positive reputations, the more likely a scientific field will be relatively fragmented, fluid, and adhocratic. Conversely, the greater their autonomy, the more control over resources is concentrated and the more audiences are highly restricted and ranked in importance, the more likely a scientific field will be highly co-ordinated and integrated, hierarchically structured and governed by highly standardized and formal rules.

4. Increasing the availability of jobs, research funds, facilities, and variety of legitimate audiences encourages intellectual differentiation and pluralism where problems and strategies are not standardized and unified.

5. Increasing the degree of concentration of control over access to key resources encourages scientists to co-ordinate their research strategies and increases the level of competition where audiences are not very diverse and equivalent.

6. Increasing the level of diversity and equivalence of legitimate audiences reduces the need for co-ordination of research strategies and encourages differentiation of topics.

7. Low levels of reputational autonomy hinder the development of highly co-ordinated and standardized research strategies and procedures.

Notes and references

[1] See, for example, the discussion of how the classical disciplinary labels in the 'life sciences' have lost their meaning in Committee on Research in

the Life Sciences of the Committee on Science and Public Policy of the National Academy of Sciences, *The Life Sciences*, National Academy of Sciences, Washington, D. C., 1970, pp. 230–9.

[2] As suggested by Randall Collins is the case for recent astronomical research, see his *Conflict Sociology*, New York: Academic Press, 1975, p. 512.

[3] As reported by Mitchell Ash, 'Wilhelm Wundt and Oswald Külpe on the Institutional Status of Psychology', in W. G. Bringmann and R. D. Tweney (eds.), *Wundt Studies* Toronto: Hogrefe, 1980. See also B. Kuklick, *The Rise of American Philosophy*, Yale University Press, 1977, pp. 459–63.

[4] Compare M. Furner, *Advocacy and Objectivity. A Crisis Professionalisation of American Social Science*, University of Kentucky Press, 1975, pp. 278–304; D. Ross, 'The Development of the Social Sciences', in A. Oleson and J. Voss (eds.), *The Organisation of Knowledge in Modern America, 1860–1920*, Johns Hopkins University Press, 1979.

[5] As in the perennial discussions of the value of operations research. See, for example, R. L. Ackoff, 'The Future of Operational Research is Past', *Journal of the Operational Research Society*, 30 (1979), 93–104; S. Eilon, 'The Role of Management Science', *Jnl. Operational Research Society*, 31 (1980), 17–28; L. G. Sprague and C. R. Sprague, "Management *Science?*' Interfaces, 7 (1976), 57–62.

[6] As discussed by E. Yoxen, 'Life as a Productive Force: Capitalising the Science and Technology of Molecular Biology', in R.M. Young and L. Levidow (eds.), *Studies in the Labour Process*, London: CSE Books, 1981.

[7] Compare the comments by Hutchison on the post-war growth of business and state demands for prediction and control strategies from economics, T. W. Hutchison, *Knowledge and Ignorance in Economics*, Oxford: Blackwell, 1977, chs. 2 and 4.

[8] The low prestige of statistics in the 1930s encouraged some statisticians to defend Rhine against attacks from psychologists according to S. Mauskopf and M.R. McVaugh, 'The Controversy over Statistics in Parapsychology 1934–1938', in S. Mauskopf (ed.), *The Reception of Unconventional Science*, Colorado, Westview, 1979.

[9] To some extent, of course, control over skills implies some control over concepts because how research is done determines, to some degree, what can be done and how it can be described.

[10] That is, they control the production of knowledge through controlling the production of knowledge production skills; see M. S. Larson, *The Rise of Professionalism*, University of California Press, 1977, ch. 2.

[11] As reflected by most of the papers in P. Abrams et al. (eds.), *Practice and Progress: British Sociology 1950–1980*, London: Allen & Unwin, 1981 and R. A. Kent, *A History of British Empirical Sociology*, Aldershot: Gower, 1982, chs. 5 and 6.

[12] See the discussion by Hutchison, op. cit., 1977, note 7, ch. 4 and Appendix.

[13] As in the recent SSRC decision to stop funding the Cambridge Economic Policy Group. See S. Weir, 'The Model that Crashed', *New Society*, 12 (August 1982), pp. 251–3 and M. Artis, 'Why do Forecasts Differ?' Bank of England paper presented to the panel of Academic Consultants, no. 17, 1982.

[14] H. Mintzberg, *The Structuring of Organisations*, Englewood Cliffs, New Jersey: Prentice-Hall, 1979, pp. 185–211.

[15] On the domination of elite groups in advisory committees and preferential treatment for a small number of departments in allocating resources in Britain, see S. S. Blume, *Toward a Political Sociology of Science*, New York: Wiley, 1974, pp. 193–212; S. S. Blume and R. Sinclair, 'Chemists in British Universities: A Study of the Reward System in Science', *American Sociological Review*, 38 (1973), 126–38; C. Farin and M. Gibbons, 'The Impact of the Science Research Council's Policy of Selectivity and Concentration on Average Levels of Research Support', *Research Policy*, 10 (1981), 202–20.

[16] As discussed by, for instance, J. Ben-David, *The Scientist's Role in Society*, Englewood Cliffs, N. J. Prentice-Hall, 1971, chs. 7 and 8; D. Kevles, 'The Physics, Mathematics and Chemistry Communities: A Comparative Analysis', in A. Oleson and J. Voss (eds.), *The Organisation of Knowledge in Modern America, 1860–1920*, Johns Hopkins University Press, 1979; P. Lundgreen, 'The Organisation of Science and Technology in France, a German Perspective', in R. Fox and G. Weisz (eds.), *The Organisation of Science and Technology in France, 1808–1914*, Cambridge University Press, 1980.

[17] T. Shinn, 'Scientific Disciplines and Organizational Specificity: the Social and Cognitive Configuration of Laboratory Activities', in N. Elias *et al.* (eds.), *Scientific Establishments and Hierarchies*, Sociology of the Sciences Yearbook 6, Dordrecht: Reidel, 1982.

[18] Although Sviedrys suggests that by the end of the nineteenth century the professors exerted a high degree of control over research strategies and approaches in physics; see R. Sviedrys, 'The Rise of Physical Science at Victorian Cambridge', *Hist. Stud. Phys. Sciences*, 2 (1970), 127–51 at p. 144.

[19] According to A. Kuper, *Anthropologists and Anthropology, the British School, 1922–1972*, Harmondsworth: Penguin, 1975, pp. 154–60.

[20] See, for instance, Lundgreen, op. cit., 1980, note 14; C. E. McClelland, *State, Society and University in Germany, 1700–1914*, Cambridge University Press, 1981, chs. 5 and 6.

[21] As reported by J. M. Beyer and T. M. Lodahl, 'A Comparative Study of Patterns of Influence in United States and English Universities', *Administrative Science Quarterly*, 21 (1976), 104–29.

[22] See, for instance, the discussion in J. Gaston, *The Reward System in British and American Science*, New York: Wiley, 1978, pp. 47–54.

[23] J. Galtung, 'Structure, Culture and Intellectual Style: an essay comparing saxonic, teutonic, gallic and nipponic approaches', *Social Science Information*, 20 (1981), 817–56.

[24] One major difference between the situation of French and German intellectuals, though, has been the much greater openness of the former to lay influences, especially the general Parisian intellectual public in the human sciences, and thus a reduction in their dependence on purely professional colleagues. See, for instance, J. L. Fabiani, *La Crise du Champ Philosophique (1880–1914)*, unpublished thesis, EHESS, 1980, ch. 4; W. R. Keylor, *Academy and Community*, Harvard University Press, 1975; C. Lemert, *French Sociology*, Columbia University Press, 1981, Introduction.

[25] P. Forman, J. Heilbron, and S. Weart, *Personnel, Funding and Productivity in Physics circa 1900*, Historical Studies in the Physical Sciences, 5 (1975), pp. 41–4.

[26] *Idem*, p. 55.

[27] For a useful discussion of the role of national differences in science, see A. Jamison, *National Components of Scientific Knowledge*, Lund: Research Policy Institute, 1982, ch. 5. On the importance of differences in physics between Britain and Germany in the 1920s, see: P. Forman, 'The Reception of a Causal Quantum Mechanics in Germany and Britain', in S. Mauskopf (ed.), *The Reception of Unconventional Science*, Colorado: Westview, 1979.

[28] See, for example, D. Kevles, op. cit., 1979, note 14, p. 161; R. H. Kargon, *Science in Victorian Manchester*, Manchester University Press, 1971, pp. 95–108; G. K. Roberts, 'The Establishment of the Royal College of Chemistry', *Historical Studies in the Physical Sciences*, 7 (1976), 437–85.

[29] See, for example, V. Descombes, *Modern French Philosophy*, Cambridge University Press, 1980, pp. 5–7; C. Lemert, op. cit., 1981, note 24; A. van den Barembussche, 'The Annales Paradigm: a case study in the growth of historical knowledge', in W. Callebaut *et al.* (eds.), *Theory of Knowledge and Science Policy*, Ghent: Communication and Cognition, 1979.

[30] As discussed, for instance, by J. Gaston, *Originality and Competition in Science*, Chicago University Press, 1973, ch. 7.

[31] Compare S. Weir, op. cit., 1982, note 13, on the allocation of funds for forecasting groups.

[32] As in the filling of sociology chairs in Britain by anthropologists, see, A. Kuper, op. cit., 1975, note 19, p. 152.

[33] As discussed by M. Ash, op. cit., 1980, note 3. See, also, Ulfied Geuter, 'The Uses of History for the Shaping of a Field: Observations on German Psychology', in L. Graham *et al*, (eds.), *Functions and Uses of Disciplinary Histories*, Sociology of the Sciences Yearbook 7, Dordrecht: Reidel, 1983.

[34] As with the private foundations in biology and the social sciences.

[35] T. W. Hutchison, op. cit., 1977, ch. 4; H. Katouzian, *Ideology and Method in Economics*, London: Macmillan, 1981, ch. 7.

[36] M. Furner, op. cit., 1975, note 4; D. Ross, op. cit., 1979, note 4. Compare P. Deane, 'The Scope and Method of Economic Science', *The Economic Journal*, 93 (1983), 1–12.

[37] For a brief review of some work in management studies, see R. Whitley,

'Management research: the study and improvement of forms of cooperation in changing socio-economic structures', in N. Roberts (ed.), *Use of Social Sciences Literature*, London: Butterworths, 1977.

[38] These points are discussed further in R. D.Whitley, The Development of Management Studies as a Fragmented Adhocracy', *Social Science Information*, 23, 1984.

[39] For some recent discussions of the various audiences for British sociology, see the papers by J. A. Barnes, A. Heath and R. Edmondson, and P. Abrams in P. Abrams *et al.* (ed.), *Practice and Progress: British Sociology 1950–1980*, London, Allen & Unwin, 1981.

[40] Such as those at Chicago, Columbia, and Harvard, see N. Mullins, *Theories and Theory Groups in Contemporary American Sociology*, New York: Harper & Row, 1973, pp. 139–40.

[41] See Kuper, op. cit., 1975, note 19, ch. 5.

[42] It should be borne in mind, though, that the degree of vertical concentration of control over access to key resources, and hence co-ordination of research strategies, is lower in many US universities than in Europe. Thus the degree of strategic dependence was not as high in 1950s US sociology as in, say, German psychology but it was higher than in post-1960s US sociology.

[43] Mullins, op. cit., 1973, note 40, pp. 139–40; E. A. Tiryakian, 'The Significance of Schools in the Development of Sociology' and N. Wiley 'The Rise and Fall of Dominating Theories in American Sociology', both in W. Snizek *et al.* (eds.), *Contemporary Issues in Theory and Research, A Metasociological Perspective*, London: Aldwych Press, 1979.

[44] Mullins, *idem*, p. 64.

[45] As reported by G. W. Stocking, 'The Scientific Reaction against Cultural Anthropology, 1917–1920', in his *Race, Culture and Evolution*, New York: Free Press, 1968. Compare R. Darnell, 'The Professionalisation of American Anthropology', *Social Science Information*, 10 (1971), 83–103.

[46] Such as Berkeley, Michigan State, Minnesota, and Wisconsin in sociology, see Mullins op. cit., 1973, note 40, p. 139.

[47] *Idem*, pp. 194–214.

[48] These departments also control the reputational system through refusing permanent appointments for intellectual deviants as happened at Harvard in the 1970s. It should, however, be noted that the 'radical political economists' were able to find academic jobs even if it was not at such a prestigious location.

[49] As reported by G. Stigler and C. Friedland in G. Stigler, *The Economist as Preacher*, University of Chicago Press, 1982, pp. 193–9 and *Doctorate Production in the United States, 1920–1962*, publication no. 1142, National Academy of Sciences National Research Council, Washington, D. C., 1963, p. 11.

[50] Stigler and Friedland, *idem*, p. 197.

[51] As in the development of different areas of artificial intelligence, see J. Fleck, 'Development and Establishment in Artificial Intelligence', in N. Elias *et al.* (eds.), *Scientific Establishments and Hierarchies*, Dordrecht: Reidel, Sociology of the Sciences Yearbook, 6. 1982, and the same author's M.Sc. dissertation entitled *The Structure and Development of Artificial Intelligence*, University of Manchester, 1978, ch. 3.

[52] See, for example, K. Knorr-Cetina, *The Manufacture of Knowledge*, Oxford: Pergamon. 1981; B. Latour and W. Woolgar, *Laboratory Life*, London: Sage, 1979.

[53] Latour and Woolgar, op. cit., 1979, p. 72.

[54] *Idem.*

[55] On the Rockefeller Foundation's influence on biological research in the 1930s, see P. Abir-Am, 'The Discourse of Power and Biological Knowledge in the 1930s', *Social Studies of Science*, 12 (1982), 341–82; R. Kohler, 'Warren Weaver and Rockefeller Foundation Programs in Molecular Biology', in N. Reingold (ed.), *The Sciences in The American Context*, Washington, D. C.: Smithsonian Institution Press, 1979; Yoxen, op. cit., 1981, note 6. On the politics of medical research funding in the USA, see S. Strickland, *Politics, Science and Dread Disease*, Harvard University Press, 1972.

[56] K. Knorr-Cetina, 'Scientific Communities or Transepistemic Arenas of Research?' *Social Studies of Science*, 12 (1982), 101–30.

[57] As discussed for Britain by, *inter alia*, Gaston, op. cit., 1978, note 22; Yoxen, op. cit., 1981, note 6.

[58] But not between mathematicians in separate, highly specialized, areas which is what I think Hagstrom and Hargens are primarily talking about when they suggest that US mathematics is 'anomic'. See W.O. Hagstrom, *The Scientific Community*, New York: Basic Books, 1965, pp. 227–35; L. L. Hargens, *Patterns of Scientific Research*, Washington, D. C.: American Sociological Association, 1975, pp. 10–16. I am also not convinced of Hargens's interpretations of the responses to his questions in relation to what he terms 'functional integration'.

[59] Compare the very brief account in R. Collins, op. cit., 1975, note 2, pp. 510–12. On the development of pure mathematics in France and Germany in the early nineteenth century see the papers by Grabiner, Scharlan, Grattan-Guinness, and Dauben in H. N. Jahnke and M. Otte (eds.), *Epistemological and Social Problems of the Sciences in the Early 19th Century*, Dordrecht: Reidel, 1981.

[60] On the role of *grands patrons* in French science see: R. Fox, 'Scientific Enterprise and the Patronage of Research in France, 1800–70', *Minerva*, 11 (1973), 442–73.

[61] As described by Shinn, op. cit., 1982, note 17 for mineral chemistry.

[62] See, for instance, B. Gustin, *The Emergence of the German Chemical Profession, 1790–1867*, University of Chicago unpublished thesis, 1975, chs. 6 and 7; Lundgreen, op. cit., 1980, note 16. On the development of chemical

research in American industry, see K. Birr, 'Industrial Research Laboratories', in N. Reingold (ed.),. *The Sciences in the American Context*, Washington D. C.: Smithsonian Institution, 1979.

[63] P. Forman, *The Environment and Practice of Atomic Physics in Weimar Germany*, unpublished PhD dissertation, University of California, Berkeley, 1967, p. 272.

[64] Lundgreen, op. cit., 1980, note 16.

[65] As reported in H. Skolnik and K. M. Reese (eds.), *A Century of Chemistry, the Role of Chemists and the American Chemical Society*, Washington, D. C.: American Chemical Society, 1976, pp. 11–64; S. Weart, 'The Physics Business in America, 1919–1940', in N. Reingold (ed.), *The Sciences in the American Context*, Washington D. C.: Smithsonian Institution Press, 1979.

[66] Forman, op. cit., 1967, note 63, p. 316; D. J. Kevles, *The Physicists*, New York: Knopf, 1978, chs. 13 and 14.

[67] See Skolnik and Reese, op. cit., 1976, note 65.

[68] But not in chemistry which still regards itself as a 'little science'. See NAS, Committee for the Survey of Chemistry, *Chemistry: Opportunities and Needs*, Washington D. C.: National Academy of Sciences, 1965. Compare S. S. Blume and R. Sinclair, *Research Environment and Performance in British University Chemistry*, London: HMSO, 1973, pp. 18–19.

[69] As exemplified by Davis's solar-neutrino experiment. See T. J. Pinch, 'Theoreticians and the Production of Experimental Anomaly: the case of solar neutrinos', in K. Knorr *et al.* (eds.), *The Social Process of Scientific Investigation*, Sociology of the Sciences Yearbook 4., Dordrecht: Reidel, 1980.

[70] On changes in the composition of the *Physical Review*, in the 1920s, see Weart, op. cit., 1979, note 65 as well as Kevles, op. cit., 1978. On more recent physics, see Gaston, op. cit., 1973, note 30.

[71] As reported by Forman, op. cit., 1967, note 63 and Kevles, op. cit., 1978, note 66.

[72] Such as hidden variables theories in quantum mechanics. See, for example, B. Harvey, 'The Effects of Social Context on the Process of Scientific Investigation: Experimental Tests of Quantum Mechanics', in K. Knorr *et al.* (eds.), *The Social Process of Scientific Investigation*, Dordrecht: Reidel, 1980; T. Pinch, 'What does a proof do if it does not prove? in E. Mendelsohn *et al.* (eds.), *The Social Production of Scientific Knowledge*, Dordrecht: Reidel, 1977.

[73] As some informants implied during interviews with some members of a highly prestigious English physics department ten years ago. This implication was not unconnected with the industrial possibilities of this work.

7

RELATIONSHIPS BETWEEN SCIENTIFIC FIELDS AND CHANGES IN THE ORGANIZATION OF THE SCIENCES

INTRODUCTION

So far I have largely focussed on the structure of scientific fields considered as distinct organizational entities in particular environments. In this concluding chapter I shall discuss relations of dependence between fields and how these form a distinct level of organization, that of science in general, as a particular system of generating and co-ordinating intellectual innovations about a broad range of topics. As mentioned in the last chapter, a major part of the environment of any scientific field is the set of immediately adjacent fields and their interconnections constitute an important component of its context. Similarly, once science has become established as a major, if not the dominant, system of knowledge production and validation, the particular location of a field within that system is a major factor in determining its autonomy, coherence, and direction. Changes in the structure and operation of science as a distinct institution producing distinct kinds of knowledges in industrialized societies affect the internal organization of scientific fields and their relative dependence upon one another so that patterns of differentiation and integration between sciences alter as the overall context of scientific research undergoes modification and development. The substantial changes in the role and organization of science which have occurred over the past 150 or so years have fundamentally altered the context in which scientific fields develop, change, and become transformed; it is this series of changes and their impact on inter-field relations which are the focus of this chapter.

Initially, I shall explore the extension of the concept of mutual dependence from scientists to scientific fields and briefly consider how we could analyse science as a distinct reputational system in which leaders of different fields attempt to convince one another, and powerful non-scientific groups, that they are central and crucial to this system such that their knowledge is highly significant and important for all fields. Second, the contextual factors which encourage different degrees of mutual dependence between fields will be examined along similar lines to these developed in Chapter 3. Thirdly, some of the changes which occurred in these factors during and after the First World War will be discussed together with their implications for the organization of the sciences. Finally, the intensification and extension of some of these changes after 1945 will be analysed, together with the development of state science policies and their consequences for the organization of scientific fields.

RELATIONS OF MUTUAL DEPENDENCE BETWEEN SCIENTIFIC FIELDS

Scientific fields can only operate as distinct reputational organizations when the prestige they are able to offer practitioners is enough to persuade them to continue participating in research and publication. This implies that science as a form of knowledge has relatively high social status and each particular field is regarded as sufficiently scientific to be able to claim control over resources for the production of its particular knowledge. For distinct fields to become established, some notion of scientific knowledge in general must have become institutionalized and accorded considerable prestige. The importance and influence of scientific fields depend, then, on some overall conception of true, or useful, or correct, knowledge which they are seen as contributing towards and in terms of which they allocate reputations and resources for intellectual products. They therefore are dependent upon general cultural ideals of scientific knowledge and, probably more importantly in the past century or so, conceptions of these ideals held by controllers of academic and similar resources who tend to be members of, or closely connected to, one or two dominant fields which are taken to represent the highest intellectual ideals. Once science had become accepted as the sources of certain – or at least more certain – knowledge about mundane matters, reputations controlled by élites in particular fields were only of value in so far as they fitted in with current views about the nature of scientific knowledge and its present manifestation in the dominant areas of research. Thus scientific fields depend upon each other to varying degrees for their own status and access to resources mediated by that status.

Variations in the degree of mutual dependence between scientific fields reflect their extent of autonomy and independence in two ways. First, fields vary in their dependence upon work in other fields to make significant contributions to their own goals. Following the discussion in Chapter 3, we could call this functional dependence. The most common form of this aspect is, perhaps, the use of technical procedures and instruments from other fields. Second, fields vary in the extent to which they adopt criteria and standards for evaluating the significance and importance of research from other fields. Reputations may be governed by norms derived from more prestigious areas in some fields rather than being largely determined by indigenous goals and criteria. This can be termed strategic dependence.

A low degree of functional dependence between scientific fields means that they rely very little upon task outcomes from one another in producing their own distinct knowledges. At the extreme they resemble Kuhnian autarchic paradigms which develop in splendid isolation from one another and undergo only endogenous changes.[1] Where functional dependence is greater, scientific fields are more reliant on the results and products from one another and use their ideas and procedures fairly frequently. If dependence is very high, a field may have few distinctive characteristics so that its boundaries and identity may become threatened. Traditional disciplines in the biological sciences, for example, appear to have lost much of their separate status since the spread of work procedures and instruments from physics and chemistry.[2] Chemistry itself has become more dependent upon the theories and techniques of physics during the twentieth century, yet it still retains a fair degree of intellectual autonomy and institutional coherence, perhaps because it was established as one of the first university disciplines[3] and retains a reputation as a relatively 'hard' science. In general, though, a high degree of functional interdependence between fields can be expected to reduce the strength of intellectual and organizational boundaries and encourage the generation of trans-disciplinary techniques and procedures.

Strategic dependence between fields implies more of an integration of goals and strategies so that they co-ordinate their approaches and problems. Where it is relatively limited, sciences are able to pursue distinct goals and objectives without concerning themselves overmuch with the significance of their reputations and their scientific status as determined by other fields. Here, scientists do not seek to demonstrate how crucial their problems and concerns are for science as a whole or to influence colleagues in other areas. A greater degree of strategic dependence implies that the relative standing of scientific fields in science as a whole is more important to scientists and they engage in more intensive competition over their centrality to scientific ideals. Problems and strategies are more co-ordinated across fields in this situation so that the output from one field can feed into and affect work in others. This implies a ranking of goals across fields so that contributions to some are regarded as more worthy and substantial than those to others. Just as major research in particle physics is more highly regarded than in solid state physics, so too work in physics as a whole is ranked above work in chemistry or biology. The goals of physics are seen as being more significant and central to current conceptions of science than those in other fields. The more this hierarchy of goals and problems is accepted in the sciences, the more scientists orient their strategies and research programmes to higher-ranked goals and seek to acquire reputations for contributions to them. Scientific fields here become more co-ordinated around some central objectives, and conceptions of scientific knowledge so that strategies are more interconnected across fields.

These two aspects of dependence between scientific fields are interconnected to the extent that a high degree of functional dependence is unlikely to occur without a fair degree of strategic dependence and vice versa. The transfer of techniques and instruments from one field to another, for instance, implies that the approach and goals they manifest are significant for the second field as well as for the first. The use of chemists' and physicists' apparatus in the biological sciences implies a certain view of scientific knowledge and ways of working which is held to be superior to those previously used.[4] Thus reputations in biology were seen as being less central to science as a whole than those in physics and chemistry. Given that techniques and results are never theory neutral, functional dependence always implies some degree of strategic dependence.

A high degree of strategic dependence similarly suggests some functional connections between scientific fields. If scientists in different areas need to demonstrate the overall significance of their work to colleagues in other fields as well as the one they are currently working in, they will be encouraged to appropriate approaches and procedures from the most central and prestigious areas in order to convince the largest and most important audience of the scientific nature of their research. The necessity of reaching a wider audience than one's immediate specialist colleagues renders cognitive boundaries permeable and liable to innovations from more significant fields. The growth of a distinct notion of scientific knowledge and the scientific method, based on the laboratory sciences, which became regarded as the generator and guarantor of truth, increased both the importance of being scientific for intellectual fields claiming legitimacy and access to material resources, and the likelihood of common approaches and techniques being used in different fields. Thus, many human sciences have sought to establish their scientific credentials by adopting methods thought to be characteristic of the laboratory sciences, especially since 1945. The prestige of chemistry and physics has diminished the appeal of alternative models of scholarship and understanding, and their associated research procedures, for many social scientists.

This process is not, though, inevitable; nor does it always occur without disputes and conflict. The importance of being scientific may have grown in the past century but what that means has changed and been subject to alternative interpretations. The use of 'hard' scientific apparatus in biological problems has not been accompanied by universal approval of physicists' approaches to these problems, nor of their intellectual priorities.[5] The greater functional dependence of the natural sciences does not necessarily mean that all biologists accept the superiority of physicists and chemists or their greater skills in resolving biological problems. Equally, although chemists are dependent upon physicists' techniques and theoretical approaches, they do not necessarily accept their priorities or legitimacy in legislating for significance criteria in chemistry.[6]

Generally, there may be considerable functional dependence between scientific fields without a similarly high degree of strategic dependence being evident, as when diverse and plural audiences and funding agencies encourage goal divergence and different significance criteria despite common techniques and instruments, which seems to characterize many bio-medical fields.[7] However, a high degree of strategic dependence between reputational organizations is likely to lead to considerable interdependence of methods, approaches, and results as fields compete over the definition of scientific knowledge and the best ways of producing and validating it. Where problems and approaches have to be considered scientific to gain legitimacy and resources, and compete for centrality to current notions of scientificity, total autonomy from other fields is improbable. In seeking general legitimacy among all sciences, or at least the more prestigious ones, scientists will tend to adopt task outcomes from their competitors to a much greater extent than if they could pursue their own goals without needing to co-ordinate research strategies across fields or demonstrate the overall significance of their problems and approaches. While high functional dependence can occur without an equally high degree of strategic dependence, then, the reverse is less likely.

INCREASING THE DEGREE OF DEPENDENCE BETWEEN SCIENTIFIC FIELDS AND CHANGES IN THEIR ORGANIZATION

The consequences of increasing the degree of mutual dependence between scientific fields can be analysed in a similar way to that between individual researchers. The more important science becomes as a distinct social organization, producing and validating particular kinds of knowledges which have relative high societal status, the less autonomy do individual fields have from dominant conceptions of science and the more important it becomes to orient goals and procedures to internally dominant groups and obtain legitimacy from them. Additionally, disciplinary social and intellectual boundaries become weaker and the more likely are fields to become ordered into a hierarchy of significance and centrality. The growing prestige of the laboratory sciences in the late nineteenth century, for instance, encouraged hierarchical models of the sciences which downgraded the traditional disciplines of botany and zoology together with their methods and preferred models of explanation.[8]

Increasing functional dependence between scientific fields implies a greater degree of specialization of topics and skills between them as scientists from different areas come to rely on each others' work to resolve problems of general interest to science. It also implies a relatively standardized symbol system for communicating results across problems and approaches so that they can be combined and integrated. Work procedures and reporting conventions therefore have to be standardized and formally structured if functional dependence is to increase between fields. For highly specialized results to be interconnected and integrated common methods and meanings must be in use across disciplinary boundaries. This, in turn, suggests that such boundaries are relatively weak and permeable in situations of high functional dependence. For scientists to use results and ideas from other fields they must be able to understand them–thus be able to 'translate' them–and adapt their problems and tasks to these novel outputs. They also, of course, must be able to adopt external procedures and notions without being penalized by disciplinary élites for using incorrect or inappropriate ideas. Thus the autonomy and power of such élites are reduced when functional dependence between fields is increased as has happened in many biological fields since the transfer of research procedures and approaches from chemistry and physics.

A high degree of functional dependence also implies a lower degree of problem specificity within disciplines as approaches and procedures from a number of fields become interconnected and organized for the resolution of particular problems. Standardized techniques and symbol systems suggest an increased mobility of skills between problems and topics so that the integration of skills and problems is weakened and scientists are no longer constrained by their skills to work on issues particular to the discipline. Skills become generalized over a wide range of problems which may cross traditional skill boundaries. Major problems now require the combination and co-ordination of specialized skills from a variety of backgrounds which are available for a variety of research topics as demonstrated by Latour and Woolgar's account of the construction of TRF.[9] Disciplinary integration of skills with problems and disciplinary control over how the latter are conceived and approached therefore decline as functional dependence between scientific fields grows.

This decline of disciplinary autonomy and coherence is paralleled by the formation of sub-groups based on specialized problems and combinations of skills organized for specific issues. These sub-groups may be established for a variety of reasons and by a variety of influences, including funding agencies, and often cross traditional boundaries. They tend to be more fluid than academic disciplines because they do not combine the control of skill production and certification with the control of jobs and reputations. Increasing functional dependence, then, encourages the development of specialist groups which have weaker boundaries and identities and are more liable to influences from outside members' fields. The degree of differentiation of the sciences into stable, firmly bounded organizational entities therefore declines. Specialization here is high but does not lead to strong stable reputational organizations being formed.[10]

Increasing strategic dependence between scientific fields, in addition, encourages competition between fields and greater efforts to demonstrate their mutual importance and significance. Research strategies become more oriented to work and priorities in other fields and scientists attempt to obtain reputations across the major fields rather than simply in their current one. The relative ordering of problems, skills, and disciplines within science as a whole becomes more critical to scientists as they compete for centrality and prestige. The more prestigious science becomes as a form of knowledge and the more it is used as a means of legitimating other skills and competences–as in the professions for example–the more crucial it becomes for intellectual disciplines to be considered scientific and the more they attempt to interpret that term in ways favourable to their own status. Equally, scientific fields attempt to distance themselves from lay criteria and audiences in favour of scientific ones to boost their prestige and legitimize both knowledge claims and resources claims as has happened in management studies since the 1950s[11]. Increasing strategic dependence thus implies greater emphasis on the demarcation of scientific knowledge, problems, and approaches, from non-scientific ones as well as more internally oriented goals and priorities.

Within scientific fields, an increased degree of strategic dependence between fields implies a greater emphasis on being scientific and competing schools using approaches and methods from other fields which are considered to be more scientific. Problems which are significant for other areas are likely to become more highly regarded as are approaches which offer benefits for scientists elsewhere. Internal disputes and conflicts therefore will become influenced by external pressures and priorities. Because reputations within science as a whole are more crucial than 'local' reputations in any single field, scientists will develop intellectual strategies which appeal to powerful groups in other fields that are more prestigious. Greater co-ordination and integration of goals and strategies across scientific fields is thus likely. This in turn may lead to a reduction in theoretical pluralism as fields become more focused on a common understanding of scientific knowledge and increasingly rely on standardized methods of producing and evaluating it. If significance criteria and priorities become science wide, rather than being largely determined by separate disciplines or specialties, and reputations ordered in terms of these criteria across fields, then substantial theoretical divergence is likely to be regarded as being non-scientific and scientists will mostly follow the dominant pattern of knowledge production and evaluation. Competition for reputations in the most central field will become intense and peripheral areas receive little attention, as will deviant conceptual approaches. A very high degree of strategic dependence between sciences, then, seems likely to produce a rather monolithic knowledge production system in which diversity, especially over conceptual and epistemological issues, is constrained and limited.

A high degree of strategic dependence between scientific fields also implies more concern with the co-ordination and integration of research from different areas just as it implies considerable concern with theoretical co-ordination of results within fields. The capacity to interrelate and generalize task outcomes across fields will be more highly rewarded in these situations than where strategic dependence is more limited. In the limit a group of 'science integrators' could perhaps emerge who perform a similar role to theoreticians in physics although this seems unlikely as long as disciplines are not organized into a single, strong hierarchy dependent upon a single source of support and legitimacy. However, attempts at systematizing the sciences into some common scheme have been made and no doubt will continue as pressures to centralize control of resources and goals increase. Once significance criteria become transscientific and the crucial reputational system becomes the organization of scientific fields considered as a totality rather than individual fields, scientists will seek to contribute to science wide goals and priorities and obtain reputations from dominant groups in the most prestigious fields. This tendency, of course, reduces further the autonomy of particular disciplinary goals and the ability of individual problem areas to set their own, unique assessment standards and intellectual priorities.

These consequences of a high degree of both functional dependence and strategic dependence between scientific fields have not yet been realized because science neither completely dominates the 'means of orientation' in industrialized countries,[12] nor is it a fully integrated and unified organization with a single identity. The contextual factors which encourage such a tight monopoly have not yet developed to this degree and not all the changes which have occurred in the structure and control of the sciences over the past century or so lead in this direction. However, in general it seems reasonable to suggest that the overall level of dependence between the sciences has increased over this period, especially in the past thirty years or so.

CONTEXTUAL CHANGE AND THE GROWTH OF MUTUAL DEPENDENCE BETWEEN SCIENTIFIC FIELDS

The major contextual factors affecting the degree of mutual dependence between scientific fields are similar to those affecting the degree of dependence between scientific competing for reputations. They can be summarized under three main headings: the extent of independence and prestige of scientific knowledge relative to competing knowledge production and validation systems, the extent to which access to jobs, facilities, and funds for producing valued knowledge is concentrated and restricted; and the plurality and diversity of audiences and sponsors for this sort of knowledge in the wider social system. The more prestigious and powerful a particular conception of knowledge, such as that typified by the laboratory based fields of chemistry and physics, becomes in a society, and the more that knowledge production is concentrated in a relatively small number of institutions, such as university departments or national academies, the greater the degree of mutual dependence between scientific fields and the more important it becomes for intellectual work to be considered 'scientific'. If, in addition, resource controllers for, and consumers of, scientific knowledge are restricted to a small number of agencies pursuing similar goals, this mutual dependence is further reinforced. Thus, the more knowledge producers are funded by central state agencies and justify themselves by being 'useful' to those agencies rather than a wider range of 'clients', the more they are constrained to produce a particular kind of knowledge and are dependent upon the currently dominant notion of science.

In very general terms, it seems reasonable to suggest that these three broad factors have changed in such a way as to increase the degree of mutual dependence between scientific fields in most countries during the nineteenth and early twentieth centuries. Natural philosophy and natural history gradually became incorporated into the European university systems and separated from 'amateur', lay participation in the nineteenth century. Laboratory-based disciplines subsequently developed a separate and superior identity to other intellectual production styles which became much appreciated for their role in two world wars and aiding profitable technological change. Science consequently became both highly valued and largely restricted to a certain kind of understanding and set of procedures for constructing it. State science policies have been developed to direct and control this knowledge production system which has expanded greatly in the past thirty years or so and broken out of the university system.

The most obvious contextual factor affecting the degree of dependence between scientific fields is the overall autonomy and influence of science as a distinct system of knowledge production. The greater this became, and the higher the prestige of reputations obtained within this production system, the more dependent are scientific fields upon their relative status within it and the more they seek to demonstrate their adherence to particular intellectual norms and procedures rather than seeking external legitimacy and direct access to resources. Thus the growth of the natural sciences as the dominant system of knowledge production, which increased our understanding of the world and provided a measure of validation for truth claims, increased the degree of mutual dependence between intellectual fields seeking legitimacy through their scientific status.

In contrast, where intellectuals had close connections with lay audiences and sought recognition from them, their mutual dependence was obviously lower and the scientificity of particular fields of relatively little concern. The traditional humanities are examples of fields which were, and to some extent still are, primarily oriented to particular lay publics rather than seeking academic respectability and 'scientific' status although this has changed somewhat with the growth in the number of practitioners competing for reputations and 'professionalism'.[13] Art historians, for instance, may use results from scientific techniques in assessing provenance but rarely attempt to legitimize their work as 'scientific'. Although lay audiences are much less important in the creation and assessment of reputations in the fields than they were in the nineteenth century, linguistic and conceptual autonomy is clearly not as great as in the natural sciences and there are few reasons to expect it to become so considerable, despite complaints about 'professionalism' in literary studies.[14] The increased degree of mutual dependence between academic practitioners in this fields since the Second World War, especially since the 1960s, has not completely cut it off from lay groups and many practitioners continue to write for general literary periodicals. To some extent, the 'literate public' is still a valid audience for academics seeking reputations.

The case of the humanities highlights the difference between academic fields of study and the modern sciences in the twentieth century. Previously combined in the universities, the humanities have largely remained there while natural science, and some of the social sciences, developed a considerable degree of institutional autonomy from academic ideals and structures. This autonomy and correlative prestige has increased the degree of dependence between scientific fields and sharpened the distinction between scientific knowledge and other forms of understanding. Whereas the natural sciences acquired institutional identity, resources, and autonomy in the nineteenth-century European universities, they have today become a separate institutional entity claiming considerable resources and a monopoly over truth production and validation. The autonomy from lay groups and standards which entrenchment in the universities produced has now been greatly augmented by the demonstration of the power of natural scientific knowledge in two world wars and in a variety of non academic programmes. Thus to be scientific in the sense established by the physical sciences is now more important than to be academic – where the latter has not become reduced to the former – and mutual dependence between scientific fields is greater than that between academic disciplines.

None the less, the growing domination of intellectual production and assessment by the universities did substantially increase the degree of dependence between fields and their separation from lay influences and standards. In addition to creating distinct intellectual labour markets which were largely controlled through reputations and certified skills, this process encouraged disciplinary competition and concern for academic respectability. Although academic disciplines formed strongly bounded social and intellectual organizations controlling the production and certification of skills, and the knowledge produced by these skills, they varied in their ablility to set their standards and control resources. Some were, and are, more important than others and were able to impose their ideals and procedures upon aspirant areas. This inequity of power is perhaps most obvious when new posts and departments are being created, or existing ones being cut and demolished,[15] but is a continuing feature of academic activity. The important point here is that fields compete with me another for local and national resources within the academic system which has a substantial degree of autonomy from other groups and institutions so that their mutual dependence is considerable.[16] Academic norms and values directed research strategies and separated knowledge production from lay, amateur efforts so that truth became a monopoly, or nearly so, of academics. Thus claims to academic status and resources became a critical aspect of intellectual development, and dominant academic ideals and conceptions of knowledge controlled intellectual work.

The relative autonomy and independence of knowledge production and validation from lay influences which academic institutionalization permitted was linked to the second major contextual factor affecting the degree of mutual dependence: its centralization in a particular type of work organization, the university. Despite the pluralism of fields organized into distinct disciplines and the derivative decentralization of control over personnel policies and resources to departments, the academic monopoly of knowledge production meant that a single institution controlled the legitimation of intellectual work and the bulk of the resources for achieving it. In this sense research became highly horizontally centralized. Such centralization encourage debates about the relative importance of intellectual fields and the nature of knowledge produced by them. Disputes over the relative merits of scientific and other forms of understanding, and the precise nature of the former, were common in the nineteenth century as different groups competed for control over intellectual ideals and institutional goals in European universities.

By organizing knowledge production into distinct units located in a particular employment structure, intellectual fields became oriented to one another to a much greater degree than if they had remained outside employment organizations or located in a variety of them. Their mutual strategic dependence was quite considerable, even in states which decentralized educational policies and institutions.[17] The separation of research in, say, geology, from lay influences and standards[18] was accompanied by its increased dependence upon other academic groupings and their ideals of knowledge production and validation. The growing horizontal centralization of control over the means of intellectual production and distribution throughout the last century thus increased the degree of mutual dependence between intellectual fields and restricted the developing social sciences to dominant academic conceptions of knowledge and forms of organization.

This increased dependence was, however, mitigated to some extent by the expansion of the university system in many countries in the early decades of this century and the organization of universities into distinct disciplines. Just as the ready availability of resources and a plurality of legitimate audiences reduces the degree of dependence within scientific fields, so too the existence of substantial resources for research in a variety of fields and the institutional differentiation of fields restricts dependence between scientific fields. Strong hierarchies of intellectual significance and centrality to organizational goals are unlikely to develop as long as resources are generally available and audiences are relatively independent of one another. If disciplines are able to control access to resources such as jobs, journals, and research funds without having to convince other fields of the overall importance of their work, they will not try to integrate their objectives and skills with those of other disciplines but simply stress their general academic respectability in a relatively diffuse manner. Once established as a distinct academic field of research and education, university disciplines are able to pursue their own purposes and certify their own skills without needing to specify how their research strategies and results fit in with other groups. Thus fields which are able to produce knowledge with few resources, and which still have some connections with non-academic audiences and general cultural élites, such as the humanities, can remain relatively mutually independent even when they are completely dependent upon university employment. In so far as such fields have some cultural legitimacy as knowledge-producing organizations and are not identical to current conceptions of natural scientific fields, they provide some alternatives to those conceptions for nascent fields such as some social sciences.

As long as the physical and biological sciences could similarly work with local and limited resources and had substantial contacts with lay, 'amateur' practitioners as in the field sciences for much of the nineteenth century,[19] they too were limited in their degree of mutual dependence. However, with the growing domination of laboratory methods and procedures for producing knowledge, extensive lay participation became less feasible and scientists restricted legitimacy and communication to colleagues employed in similar organizations and with similar facilities. By limiting possible contributors to those with access to increasingly expensive equipment located in a relatively small number of research centres, leaders of laboratory-based fields could control training, certification and knowledge production to a much greater degree than those in sciences which relied upon extensive observations over a wide variety of conditions and circumstances. Dependence within each scientific field thus increased, and as the increased cost led to reliance on a few state or private agencies so too did dependence between the natural sciences.

An additional contextual influences on the degree of mutual dependence between sciences is the sheer number of fields claiming to be scientific and competing for legitimacy to gain access to scarce funds and facilities. Just as an increase in the number of researches in a particular scientific field often leads to an increase in the degree of mutual dependence between them and greater specialization, so too the more disciplines or fields claiming scientific status, the more dependent they are upon one another and dominant ideas of scientific knowledge, skills, and work procedures. This dependence includes both functional and strategic aspects because competition encourages specialization of topics and approaches within some overall characterization of scientific knowledge and fields attempt to ensure that their outputs fit in with one another and are useful to work elsewhere, especially when resources are limited. They therefore tend to standardize work methods and symbol systems so that results and techniques can be transferred and thus shown to be relevant across traditional boundaries.

This transfer of instruments and approaches between scientific fields is also aided by internal pluralism and conflicts within them, as well as by their being ordered in terms of important and degree of scientificity. The less prestigious a field is, and the more it is divided into competing schools which hold divergent conceptions of its subject matter and appropriate ways of dealing with it, the more open it will be to techniques and analytical methods from more prestigious and central fields. Or, at least it becomes more likely that some groups will seek to boost their own reputations and control over the direction of research by adopting currently fashionable procedures and styles of reporting results and so claim greater 'scientific' status. This is perhaps most clearly seen in the use of mathematics in the human sciences but also occurs in the use of metaphors and analogies from physics and chemistry in a number of fields. Functional dependence between sciences is thus increased when scientific fields are weakly integrated and bounded and highly dependent upon their scientific status for legitimacy and access to research resources.

In summary, the degree of mutual dependence between scientific fields is affected by their autonomy and independence from non-scientific groups, relative prestige and distinctiveness of science as a particular way of organizing knowledge production, the degree of concentration of control of access to the means of intellectual production, and the extent to which the major audiences for scientific knowledge are internal and hierarchically ordered; as opposed to both including lay groups and not being strongly arranged into a hierarchy of importance and centrality. During the present century these contextual variables have generally increased the degree of mutual dependence between scientific fields although the wide availability of resources for public scientific research after the Second World War has mitigated their consequences until quite recently. In the next section I will briefly discuss some of the major changes in the context of scientific knowledge production during the inter-war period and will focus on the post-1945 system in the subsequent section.

CHANGES IN THE CONTEXT OF SCIENTIFIC RESEARCH BETWEEN THE WARS

The growth of state and industrial employment organizations for scientific research in the 1920s and 1930s both emphasized the differences between the natural sciences and other fields entrenched in universities, and reduced the importance of departmental boundaries and reputational identities based upon university-circumscribed disciplines. It therefore increased the degree of mutual dependence between these scientific fields. However, by increasing the diversity of funding sources, of employment opportunities, and research goals, this growth reduced the dependence of natural scientists upon the academic establishments and so the academic conception of scientific ideals and truth criteria. Academic monopolization of control over the designation of a field as 'scientific' was therefore restricted and strategic dependence between all the sciences reduced. Although academic élites controlled the skills required for these new employers, and the integration of research results into training programmes, they no longer monopolized the purposes for which practitioners were employed and thus the level of demand for particular skills. Because this demand affected the distribution of resources within universities,[20] research goals began to be shared among university scientists and the determinants of the wider market for scientific skills. Intellectual priorities became less totally controlled by purely academic interests and more open to resource allocation decisions in other employment structures.[21] Even if academics still dominated the reputational systems of the sciences, their claims to, and control over, the increasingly expensive laboratory facilities required for the highest reputations became mediated by non-academic interests and goals. The natural sciences thus developed a distinct identity and legitimacy but this involved increasing interaction with non-academic groups and a certain consonance with their purposes and structures.

This growth in employment opportunities and in the prestige of the laboratory sciences was especially marked in the 1920s, partly because of the usefulness of chemistry and physics in the First World War.[22] The number of chemists and physicists trained in the United States, in particular, showed a spectacular rise after 1918[23] and many of these found jobs in industrial and state laboratories. According to Weart,[24] the number of research workers in US industry increased about fivefold in the 1920s, while that of physicists in industry roughly doubled and the expansion of teaching posts in physics in universities was clearly dependent upon the growth of industrial jobs for physics PhDs. Despite the severe set-backs in industrial research during the 1930s, by the end of that decade the production and industrial recruitment of physics PhDs had resumed its upward path and the number of PhDs awarded in chemistry between 1931 and 1940 was three times that of the previous decade. By 1940 the chemical industry in the USA accounted for one-fifth of total research employment.[25]

This period also witnessed a substantial change in the way research was supported and controlled. The growth of external funding for particular research projects and programmes by the Rockefeller foundation and other philanthropic organizations in the USA and Europe enabled a relatively small group of officials and their advisers to influence directly the production of knowledge in favoured areas – or of a particular type – rather than leaving control over production to university structures and preferences and the reputational system. By funding research in particular areas directly, intellectual priorities could be altered and co-ordinated by a central group which acted across universities. Although this growth was not as massive as the expansion of state funding of research in the 1950s and 1960s, it tended to set the pattern for the future organization of research in the natural sciences –

and thence for the human sciences – and enabled particular areas of interest to become the focus of disproportionate support and encouragement. Forman, for instance, suggests that atomic physics was favoured over other areas of physics, and physics over chemistry and other fields, by the *Notgemeinschaft* in the early 1920s to the extent that the severe inflation of that period had remarkably little effect on the output of papers in this areas.[26] Similarly, Kevles states 'Public or private, the institutions favoured by the concentrated distribution of the new funds harbored the peak departments of physics. Their administrators won their share of the new money for science, and year after year in the 1920s academic physics grew far better off financially than it has ever been before'.[27] The new NRC post-doctoral fellowships enabled US theoretical physics to blossom and form a national élite for physics.[28] While universities had to balance claims from competing disciplinary leaders so that new fields were unlikely to be favoured over established ones, the new funding agencies could and did pursue more directive and narrowly focused policies to achieve definite and visible results. This change was exemplified, of course, by Warren Weavers's concentration of funds in particular types of biological research so that chemical and physical techniques became dominant there.[29]

This growth of external funding of research and positions in the natural sciences not only enabled scientists from these areas to escape from the boundaries and constraints of universities and so increased their flexibility and autonomy from local goals; it also enabled the reputational élite to control directly major resources on a national scale. By centralizing control over substantial funds in a single, or small number of, agencies which could be dominated by a small group of eminent advisors, this change increased the potential control that could be exercised by dominant groups in fields where procedures and symbols were standardized and there was considerable agreement over intellectual priorities. Thus research strategies could be more integrated and co-ordinated across employment organizations through resources allocation decisions as well as through the reputational system. In physics especially, national and international policies could be pursued through these central agencies so that atomic, nuclear, and theoretical physics dominated the field and attracted the most promising graduates in the inter-war years.[30] This was less obvious for chemistry, perhaps because of the greater diversity of funding sources, including industry, and the less expensive apparatus needed there. In biology, disciplinary élites had less control over foundation officials such as Weaver who were able to pursue alternative goals and strategies and construct a new research system modelled upon physics. Centralization here was less complete and, although the rise of molecular biology represents an attempt to establish a central field, which would control research resources and priorities throughout the biological sciences, it has scarcely reached the same dominant position as elementary particle physics. Nor is there any indication that a theoretical biology will emerge to fulfill the same role as theoretical physics.[31] The case of molecular biology does indicate, though, how non-academic agencies can structure the development of new fields and alter disciplinary priorities, especially in sciences which do not have the greatest prestige and power in the science system.

The considerable growth in employment for scientific researchers, especially chemists and physicists, in the inter-war period, plus the expansion of resources for academic natural science, heightened the domination of laboratory-based science over other intellectual endeavours and other types of knowledge production. Notwithstanding the insecurity of physicists in Weimar Germany,[32] and the reaction against technological change and the physical sciences in the Anglo-Saxon countries in the 1930s,[33] these fields had become the dominant ones by the Second World War. They were seen as producing applicable and useful knowledge for a variety of goals and could be organized and controlled outside the university system by business, the state and philanthropic agencies. Furthermore, substantial investments could be, and were, made in seemingly recondite areas of university research for long-term benefits in a way which would have seemed improbable, if not ludicrous, before 1914. Nuclear physics, for example, obtained money for cyclotrons on the grounds of their medical applications in treating cancer.[34] Thus laboratory science produced not only the most precise and true sorts of knowledge, but also the basis for technical control and use of the environment and the human body. In the case of physics, the most arcane and irrelevant topics became capable of generating knowledge which could be applied to non-academic purpose so that a happy marriage of pure knowledge and useful results seemed to have become established. Of particular importance in this regard was the development of analytical and approaches which could be used to advance other, less precise, fields for the benefit of all. This 'technology transfer'[35] from physics to chemistry and biology demonstrated the legitimacy of physics' domination of science and the usefulness of external agencies in aiding the development of new areas. It also highlighted the new way of organizing and controlling public science.

Yoxen has suggested that Weaver's strategy in developing what became molecular biology exemplified a new system of patronage and control of research which encouraged 'new ways of specialized, collaborative working' on programmes formulated by funding agencies and their close advisors.[36] Whether Weaver was actually concerned to establish a new field of transdisciplinary research rather than simply to make biology more 'scientific' through technology transfer[37] need not concern us here as much as the undoubted change in the dominant pattern of the organization and control of research his programmes reflects. This change involved the development and implementation of national and international research strategies for particular fields by relatively small groups of élite scientists and agency officials who were able to direct scarce resources for specific goals and restructure the reputational system around the new priorities. Rather than research being largely a local affair organized in relatively separate organizations by research directors following their own approach and goals as in the traditional German system,[38] knowledge became seen as an object which could be directed and controlled centrally through the resource allocation system which concentrated its funds in a small number of institutions working on specialized parts of a co-ordinated programme.[39] To the extent that research leaders were able to impose their priorities upon their colleagues, and these necessitated the use of expensive facilities which could only be obtained through the funding agencies, the reputational system became ordered around the resources allocation system and highly centralized.

The use of the National Research Council post-doctoral fellowships in physics in the USA to create an élite group in quantum mechanics based in a few highly privileged universities illustrates this process.[40] Research in this field became organized by the reputational élite which controlled the training and recruitment of their successors through their control of foundation funds. By providing travel money, freedom from teaching duties, and other resources, these fellowships – which were highly sought after – enabled the putative future élite to concentrate on producing highly valued knowledge in a few select organizations and thus outdistance their age-group colleagues in the reputational system. The current élite were thereby able to control both research priorities, and the selection of their successors, at the expense of local contingencies and preferences. In this way, the Rockefeller foundation policy of 'making the peaks higher'[41] by concentrating funds in a few select universities and institutes was extended into the control of individual priorities and élite succession strategies. Thus Weaver not only encouraged the application of physicists' and chemists' techiques to biological problems, and so the reorientation of biologists' priorities, but he also attempted to transfer the physicists' model of work organization and control to biology. While this attempt was not successful before the Second World War, it heralded many aspects of the post-war organization of research in the natural sciences which subsequently became the model for many of the human sciences.

In summary, then, the inter-war years witnessed the emergence of the laboratory-based natural sciences as the dominant form of knowledge production and truth ideal and their emancipation from the university system of employment in a novel structure of research organization and control. The sciences became distinguished from other intellectual fields in universities both as producers of knowledge production skills for a variety of goals, and as systems of knowledge generation which could be directed and managed for social purposes. A new structure of organizing and controlling research on a national scale became established alongside the academic system which reduced local influences upon research strategies and intellectual goals in favour of national and international ones. This change reduced scientists' dependence upon purely academic ideals and norms, but increased their susceptibility to control by national élites and resources controllers. In so far as science became a distinct, prestigious system of knowledge generation controlled by national establishments and funding agencies, the overall degree of strategic and functional dependence increased, but where resources were not centrally controlled or concentrated in one establishment and disciplinary élites did not become allied to agency officials this increase was considerably mitigated. As long as research could largely be conducted with locally available resources and reputational boundaries could be preserved against incursions from agency officials such as Weaver, mutual dependence was limited.

THE POST-WAR ORGANIZATION OF THE SCIENCES

The prestige of the laboratory sciences, and especially physics, grew enormously after their successes in the 'physicists' war'. So too did the importance of extra-local research funding agencies, their domination by the state in most western countries and the amounts invested in scientific research for military, commercial, and social purposes. These sciences have become large-scale consumers of resources in a wide variety of settings, producing knowledge for a wide variety of goals and the object of state planning and direction in many countries. From being part of general intellectual endeavours located in cultural institutions they have developed an independent status as systems of knowledge production and producers of knowledge production skills which form an integral part of national resources and capabilities. Scientific research in general is increasingly an activity to be invested in, directed and organized on a national and international level.[42] Science has become seen as a distinct system of knowledge production and integration which can be the focus of policies and strategies within and between nation states.

In many ways these changes are simply the extension and intensification of the processes already discussed above. However, the sheer size of the resources allocated to scientific research and the number of people making a living from this activity have become so great that the science system since the 1950s cannot be adequately understood as simply 'more of the same' as the inter–war years.[43] Furthermore, the direct involvement of the state in a number of guises has led to to substantial changes in organization and control of research and the establishment of national and international science policies on a wide scale.

The exponential rise in the number of natural scientists, journals, and papers has been frequently commented upon.[44] Less noted have been the consequences of this rise for the dominant pattern of research organization and control. In particular, the increasing separation of skill production and certification from knowledge production and validation has not often been discussed nor have its implications for the traditional conception of the scientific community been systematically examined. Growth in numbers of competing researchers has usually been seen as leading to greater specialization, and hence the replacement of disciplines by specialisms as the main locus of intellectual organization and control,[45] but these are still commonly viewed as integrated, cohesive communities and little direct cognizance seems to have been taken of other changes in research organisation.

In addition to the separation of skill training and certification from much knowledge production, the most important of these changes are the increasing dependence on external funding of programmes and projects in most scientific fields, the centralization of this funding in a few state-controlled agencies, and the increasing multiplicity of goals being pursued by employers of researchers. The hybrid system of research training, employment, and validation through reputations which developed in the nineteenth century, when scientists became employed and rewarded for their contributions to knowledge as evaluated by international reputational groups, has now become much more differentiated, and a fourth component has been added: external funding agencies which directly influence research strategies. The incorporation of the natural sciences into the 'high culture' system of the reformed European and North American universities and similar institutions has now been surpassed by their establishment in a new set of institutional arrangements which increasingly form the model for other intellectual activities and structures. The universities are still important components of these new arrangements but no longer dominate them. Furthermore, their own internal structures and control procedures have altered considerably since the Second World War with the growth of large scale external funding and research entrepreneurs such as 'Dr Grant Swinger'.[46]

The growing importance of external funding for research in many sciences and its concentration in a relatively small number of state agencies reduced the autonomy of academics and universities in setting research strategies and significance criteria where existing reputational élites were able to control these agencies. By controlling the bulk of resources for producing knowledge such élites could influence research goals directly as well as controlling reputations for work already completed. Where, then, reputational organizations were already fairly centralized, as in physics, and the dominant groups controlled the criteria for allocating state funds, increased dependence on external resources heightened the degree of centralization and mutual dependence. The generalization of this pattern of organization and control to other fields by extending and expanding post-doctoral fellowship programmes, increasing the complexity and expense of research apparatus and greatly expanding state support for university research could have resulted in many sciences becoming similar to physics as a system of knowledge production and control.

To some extent this has occurred in part of the biological sciences where Weaver's efforts to change the biological sciences through encouraging the transfer of techniques from chemistry and physics, the creation of an international élite through travel grants, research fellowships, and contracts, and the co-ordination of research strategies have been echoed and augmented by state agencies since the war.[47] However, the overall consequences for the sciences of the massive growth in state funding, and attempts to direct and co-ordinate research goals centrally, have not quite followed the example of physics. Instead, the traditional patterns of integration and control through academic disciplines seem to have broken down in many fields without any coherent and stable structure emerging to replace them. The four major components of the knowledge production system – training, employment, reputational assessment, and funding – have become more highly differentiated and often oriented to different objectives so that co-ordination of research strategies within and between specialized fields is limited.

The limited extent to which most sciences have become as centralized and co-ordinated as much of physics is due partly to their lower initial degree of intellectual and organizational centralization, partly to the relative ease with which the necessary research resources could be acquired until recently because of the general expansion of funding and educational posts, but most of all because the objectives and priorities of many funding agencies did not coincide with those of existing reputational élites, were often divergent between themselves, and some times sought to develop alternative ways of producing knowledge, as in the cases of molecular biology and behaviouralist human science.[48] As a result, existing reputational leaders were unable to dominate the allocation of a large proportion of research resources through reputations in many fields and scientists could pursue a variety of research strategies without having to co-ordinate them or demonstrate the general value of their work for the field a a whole. Existing boundaries and objectives of reputational organizations became weakened and scientific fields became the product of a mixture of influences and ideals derived from traditional disciplines and their skills, employers' goals, funding agency objectives, and reputational groups in adjacent areas.

Where, then, agency goals were different from existing reputational ones, where the resources allocated to those goals were relatively massive, and where employers were oriented to largely reputational means of controlling research output, the diversity of goals and audiences in sciences has increased. Strategic dependence between fields in this situation is fairly limited and there is little need to co-ordinate and integrate strategies across problem areas. The overall significance of particular problems often cannot be decided in an unambiguous manner because of the plurality of criteria and goals held by different audiences. Thus the degree of control over research strategies exercised by any single reputational group is limited and scientists are subject to a multiplicity of influences and criteria. The separation of skills from research goals and long-term intellectual programmes enables employers and state agencies to impinge directly upon problem selection and individuals' strategies because strong reputational organizations combining all the major aspects of research work have become weakened and fragmented by their limited control over the allocation of the bulk of research funds and jobs. By directly funding research in separate organizations for distinct purposes and controlling much of the external funds needed for university research, the establishment of state agencies pursuing a variety of goals in the bio-medical sciences has not only severely reduced the power of disciplines such as physiology and zoology but also rendered the formation of any new reputational organization which could integrate and co-ordinate research around distinct intellectual goals increasingly improbable.

As a result of this dissolution of strong reputational organizations combining control over recruitment, training, jobs, and research resources, strategic uncertainty in existing reputational groups has increased considerably and scientists have no single dominant audience as the focus of their research strategies. The diffuseness of agency goals and difficulties in interpreting them for research planning purposes in many bio-medical sciences have allowed laboratory directors and officials considerable latitude in allocating resources and determining priorities. They have also made the evaluation of research outputs in terms of agency goals an uncertain and potentially highly contentious activity. The rapid expansion of the National Institutes of Health in the USA and similar organizations elsewhere enable scientists to 'shop around' for financial support and obtain funds with relative ease from other groups when rejected by their first choice. Given the problems of assessing the worth of projects and results for diffuse purposes and the general availability of resources from a number of agencies, systematic co-ordination of projects around detailed programmes has been difficult to accomplish and a wide variety of approaches have been supported.[49] The current reductions in research support and demands for public accountability are encouraging agency officials to formulate programmes and try to integrate their research efforts, but competing views about priorities and the vagueness of institutional goals in scientific terms limit their success. The opposing beliefs and interests of the medical profession and reductionist 'basic science' in West Germany have, for example, restricted the development of a coherent cancer research policy leading to a situation of 'pluralism without organization'.[50] Consequently, research tends to be organized around relatively short-term problems which can be solved without too much difficulty and a premium is placed upon technical competence in the absence of generally accepted significance criteria. In effect both researchers and administrators deal with uncertainty and conflict by reducing risks and investing in a wide variety of projects and topics which promise some definite, if limited, outcome. The difficulties of developing and implementing a systematic research policy in these fields have resulted in resources being allocated by scientists largely on the grounds of competence and thus encouraging specialization and fragmentation. Because the traditional disciplines no longer control research funds and facilities as much as they used to, and alternative systems of employment and funding have become established, intellectual coherence and integration around a few central theoretical gols has declined as scientists pursued innovation and autonomy by specializing in narrow topics and materials with standard skills and competences. The old bases of co-ordination and control have been weakened by the colonization of biology by chemists' and physicists' techniques, and by the growth in external funding, without alternative structures being able to take over their role.

Furthermore, the growth of state laboratories oriented to semi- or non-academic goals itself creates new skills and competences as scientists from different backgrounds are encouraged to work together on new topics and problems which cross traditional boundaries. Simply bringing skills and interests together for cross-disciplinary purposes redirects attention and alters perceptions of priorities. Even if scientists remain primarily oriented to their parent discipline for obtaining reputations so that they attempt to redefine research goals along those lines, they have to collaborate with others from different backgrounds and so cannot simply pursue traditional goals. Organizational constraints and structures limit the extent to which scientists are able to appropriate employers' goals and so research strategies reflect compromises. What is organizationally feasible affects research tasks considerably, as recent studies have shown,[51] and given the dominant orientation to reputational systems of rewards and recognition in many state laboratories, scientists attempt to justify and claim intellectual significance for their work by showing how important it is for others. Thus they are led to reorder intellectual priorities to suit their own organizational realities; where existing disciplinary structures are pluralistic, not central to scientific ideals and unable to command the bulk of resources this reordering has a major impact upon reputational significance criteria and norms.

To summarize, the expansion of the research system and its orientation to a variety of goals since the war have led to a proliferation of problem areas which are rarely highly connected and integrated at the strategic level but which often use similar skills and techniques. Specialization has been encouraged by the large increase in numbers of scientists competing for reputations and the domination of laboratory techniques which has provided the basis for some skill standardization across fields. Thus scientists narrowed their topics and foci to avoid direct competition yet have been able to claim contributions to intellectual goals because of common technical procedures and increasing standardization of materials. Integrating these goals into higher-order objectives has not been necessary because of the relative abundance of resources until recently and the diversity of interested parties such as funding agencies, employers, and training structures.

THE DEVELOPMENT OF STATE SCIENCE POLICIES

Bureaucratic co-ordination of agency goals and resources around some central set of objectives, however, has become increasingly important in many countries as the size and the cost of state-supported research mushroomed after the war and science became seen more as a factor of production than just as one component of 'high culture'. The development of state policies for, and through, science has now become an institutionalized activity for most national bureaucracies.[52] These policies objectify science as a national resource to be planned, managed, and improved both for intellectual and political objectives; they combine a concern for the quality of public scientific knowledge and skills with efforts to direct research towards economically and socially useful goals. Implicitly or explicitly they seek a consonance between intellectual objectives and criteria of significance and useful outcomes. In so far as these policies are accepted and implemented in a systematic way, they organize research around national priorities, demarcate science from other intellectual products, and encourage the co-ordination of goals and strategies within and between scientific fields. Such co-ordination implies a unitary conception of scientific knowledge and appropriate skills for producing it so that a uniform and administratively rational system of work organization and control can be developed. Fundamental differences between types of knowledge and ways of producing them are neglected in favour of greater standardization of research skills which can be inculcated through the educational system and applied to a wide range of problems to produce reliable and useful knowledge.

Such attempts to direct research for military or economic purposes are not, of course, an entirely post-war phenomenon. At least since the seventeenth century monarchs and others have subsidized academies, expeditions, observatories, and other research facilities for a variety of goals and the extensive reforms of state educational systems in the nineteenth century had major implications for the production and control of research as we have seen.[53] However, the systematic pursuit of political objectives by organizing scientific research on a continuing, long-term basis is relatively recent and its institutional embodiment in state bureaucracies charged with developing polices for the national management of research has only become a reality since 1945.

These policies have often involved the advice and preferences of relatively small group of élite scientists, especially physicists, who move easily between the universities, state agencies, and advisory groups.[54] Claiming competence over a wide range of activities, they form a distinct 'establishment'[55] which mediates the demands of state agencies and politicians on science on the one hand, and the claims for resources and autonomy from scientists on the other. [56] Operating across the sciences, they claim to speak on behalf of their research colleagues and to interpret general state objectives and policies for resources allocation criteria between and with scientific fields. By seeking to integrate scientific research with state policies, or at least ensuring that they are commensurable, this establishment both legitimates and sells science and tries to manage intellectual priorities. The more it monopolizes the mediation function and becomes involved in coordination of the policies and practices of funding agencies, the more it dominates inter-field relations and the more interdependent scientific fields become. Given the dominant position of physicists in such establishments, it is not surprising that is their conception of scientific knowledge, and the appropriate administrative arrangements for organizing its production and assessment which tends to dominate science policy-making. Thus priorities and ideals become focused upon a certain type of knowledge and related procedures for producing it.

The influence and prestige of this establishment, and the demonstrable utility of the knowledge it produced during the war, meant that state-supported and controlled research formed part of the public science system rather than being separated from it. Thus, scientists published their results and acquired reputations from colleagues working on similar problems elsewhere on the basis of their contributions to intellectual goals. These reputations have been the major means of allocating rewards and further research funds in many fields. Within general and sometimes rather diffuse objectives, then, scientists pursue reputations and decide on the significance of results by assessing their contributions to reputational goals. Implicitly or explicitly it is usually assumed that these goals are consonant with, and contributing to, the general mission of the employment organization and funding agencies. The long-term goals of curing cancer or building a commercially viable fusion reactor are, in effect, elaborated and applied by biologists, biochemists, and plasma physicists who seek reputations from fellow specialists in terms of their intellectual goals and priorities. By delegating much control over goals and processes to scientists pursuing reputations, funding agencies and government departments have substantially expanded the public science system and also gained a considerable degree of influence over it. State science policies increasingly affect the framework in which present reputational organizations operate and new ones develop and become established.

Four major consequences of science policy-making and implementation for the organization of scientific fields can, then, be identified. First, in so far as they increase the diversity and plurality of funding sources, criteria, and goals, they will reduce individuals' dependence upon existing reputational structures and norms. If this increase is considerable, such autonomy will lead to a reorganization of scientific fields and a lower degree of intellectual co-ordination between research strategies in different fields as we have seen in the life sciences and as can also be observed in many of the human science.

Second, in so far as they seek to organize scientific research for specific objectives which require reliable, reproducible and visible results they encourage standardization of technical procedures and a narrowing of research foci around restricted topics. The use of particular skills for various purposes decided by agency officials presumes the generalizability of those skills and their applicability to a wide range of problems. This puts a premium upon work procedures which are generalizable and not too tied to particular problems and intellectual goals. These seem to be most common in the physical sciences where phenomena are tightly restricted[57] and skills applicable to a wide range of problems. Thus the pursuit of reliable results for specific objectives tends to encourage the development of methods and approaches typical of the 'restricted' sciences.

Third, the development of a policy for scientific research

itself tends to promote a particular concept of science and how it is to produce knowledge. Concious reflection and construction of any state policy implies some view as to the nature of the object being planned. Thus in the case of science the simple consideration of a research policy reifies a particular conception of knowledge and appropriate methods and people for producing it. In the USA for instance the human sciences were omitted from the range of fields covered by the National Science Foundation when it was intitially esablished because they did not fit the dominant conception of 'science' and were regarded as controversial.[58] The growing importance of state funding and resource control in many scientific fields has increased the significance of this image so that science has become strongly associated with the sort of knowledge generated and validated in physics and especially with the technical procedures and symbolic apparatus of physics. This has been particularly noticeable, of course, in the human sciences.

Finally, state science policies encourage the generalization of particular administrative procedures for managing and directing research. The success of the war-time arrangements and apparent ability of private business to manage research for profit encouraged the belief that public science could be 'steered' in a similar sort of way to produce nuclear energy, atomic weapons, and other desired products. The systematic organization and administration of research for public goals is most visible in the large state-dependent bureaucracies established for research into atomic energy and in the national and international laboratories for high energy physics. It also occurs, though, in much of astronomy and increasingly in the bio-medical sciences where separate, large employment organizations have been established to pursue research for medical purposes. The European Molecular Biology Organization is a fairly recent attempt to deploy the 'big science' model of research planning and administration based on physics into the biological sciences.[59]

This extension and elaboration of particular administrative structures for the direction and management of research from particular fields and circumstances to sciences in general encourages the production and preferential treatment of knowledge which fits in with them. Extensive divison of labour, specialization of tasks and skills, standardization of many processes and work procedures, and elaborate mechanisms for co-ordinating task outcomes and research strategies are appropriate for, and result in, the production of a particular kind of knowledge. This sort of knowledge restricts the variety of properties of phenomena which are relevant, tends not to be very concerned with the organizational properties of objects or specific details of individual configurations, and emphasizes generalizable attributes and simple relations between large numbers of similarly constructed objects.

Even if such bureaucratic arrangements become largely organization 'fictions', as they seem to be in some bio-medical laboratories,[60] and so do not actually direct research in highly restricted ways, their existence acts as a powerful constraint on the sort of work which is done and how it is evaluated. As the dominant model for organizing scientific research it limits the development of alternative types of knowledge and encourages those whose strategies are commensurate with it. With increased demands for accountability and objective, relevant performance appraisal systems, this encouragement will grow so that funding may become much more dependent on adopting appropriate administrative arrangements and producing visible, reliable pieces of knowledge at regular, short intervals. This is likely to have especially strong consequences for the biological and human sciences where some approaches do not easily lend themselves to piecemeal production. The extension of particular organizational structures predicated upon experiences in physics to these fields threatens to reduce intellectual pluralism and diversity and encourage reductionist strategies.

More generally, these changes in the organization and control of knowledge production constitute a major alteration in the research system. The rapid growth of the scientific labour force over the past thirty years in many countries, and the increasing state support of knowledge production, have reduced the independence and autonomy of university disciplines as the dominant units of intellectual co-ordination and control so that the inculcation and certification of research skills has become increasingly separated from their employment and direction. Large parts of the enlarged system of public intellectual production and validation are developing characteristics more similar to the professions and crafts than to the traditional academic model of research. Public scientific knowledge is becoming the product of bureaucratically organized combinations of standardized skills for a variety of intellectual and non-intellectual goals. It is thus beginning to acquire some of the features of industrial, private science. The 'industrialization' or intellectual work which began in the nineteenth-century universities, and was typified by Liebig's 'knowledge factory', has developed into another phase since the Second World War in which reputational goals and values are established and changed by shifting coalitions of intellectual élites, state functionaries, and administrative leaders in employment organizations.

SUMMARY

The points made in this chapter can be summarized in the following way.

1. Relations between scientific fields can be usefully discussed in terms of their degree of mutual dependence and changes in this dimension. Two distinct aspects of mutual dependence between sciences can be distinguised: functional dependence and strategic dependence.

2. Increasing the degree of mutual dependence between scientific fields is associated with certain changes in their organization: (a) a greater self-consciousness of being 'scientific' and rejection of 'non-scientific' procedures and ideals, (b) a decline in their openness to lay standards and contributions, (c) increased concern with their relative centrality to scientific ideals and values and consequent influence upon other fields, (d) increased specificity and narrowness of their goals and problems, (e) weakened boundaries of scientific fields and increased mobility of skills and ideas between them, (f) increased co-ordination and interrelation of research goals and results across fields, and (g) the emergence of a common, 'scientific' way of

conducting research, communicating task outcomes, and organizing work throughout the sciences.

3. The degree of mutual dependence between scientific fields varies in relation to particular contextual factors. It increases when: *(a)* science becomes more prestigious and gains control over the criteria for allocating major resources such as jobs and facilities, *(b)* one particular set of intellectual values and procedures dominates the scientific prestige system and is accorded general social primacy as the most useful and valid form of knowledge, *(c)* access to substantial resources is controlled by an élite group of advisers and policy-makers drawn from the dominant field(s), *(d)* support for scientific research is dominated by a small number of agencies rather than by a diversity of audiences and groups, and *(e)* status among the sciences is more critical to gaining access to resources and social legitimacy than direct appeals to lay groups and clients.

4. As intellectual production and validation has become more dominated by university employees, and by the laboratory-based sciences, scientific fields have become more independent and separate from lay goals and standards and more internally specialized.

5. As the system of knowledge production has expanded and transcended the university structure, the inculcation and certification of research skills have become differentiated from the employment and direction of researchers so that scientific fields are no longer coterminous with academic disciplines and intellectual goals are less determined by purely academic considerations.

6. The growth of state funding of research and state policies for the direction and management of research encouraged the domination of a particular kind of knowledge production and of its organization and control. Increasingly this has become identified with the laboratory-based sciences where standardized techniques and work procedures enable researchers to produce reliable, predictable knowledge for a variety of social purposes about a variety of subjects. The more co-ordination and central planning of scientific research there is, the more similar and comparable

scientific fields will become in their dominant pattern of work organization and control and in their intellectual ideals.

CONCLUDING REMARKS

In this book I have outlined a framework for the analysis and systematic comparison of scientific fields in changing cir cumstances as a means of understanding how and why systems of intellectual production vary and alter. By focusing on the sciences as particular kinds of work organizations, I was able to identify two major dimensions according to which scientific fields vary and produce different types of knowledge. These variations, in turn, are connected to differences in certain contextual factors, and changes in these latter aspects lead to changes in the organization of particular sciences. Thus I have tried to link environmental changes and developments to changes in the intellectual and social organization of scientific fields. These links seem important to me if we are to grasp the diversity of the sciences and patterns of change in them over the past two centuries.

It is this diversity and change which are the main concerns in this book. By dealing with variations between the sciences and their particular circumstances I have suggested a way of analysing the major differences between scientific fields, providing some reasons for these differences and exploring the consequences of environmental changes. In so far as the present public sciences are organized as reputational systems they are still similar to the 'natural philosophy' and 'natural history' of the seventeenth and eighteenth centuries, and can be compared with them. However, the major changes which occurred during the nineteenth and early twentieth centuries have had substantial effects upon the organization and control of knowledge production, as I have pointed out in Chapters 2 and 7, and scientific fields are not the same phenomena today as they were then. In particular, the domination of knowledge production and certification by employees of a certain type of organization, and the later development of national support and direction of research efforts in the public sciences, have institutionalized intellectual disciplines as labour markets and subsequently combined reputational means of directing research with central control over the production of knowledge by funding agencies. These changes first sharpened intellectual and social boundaries, while increasing internal cohesion and dependence, and then encouraged the separation of reputational organizations from training organizations and of science from other intellectual endeavours, which allowed knowledge production to be steered towards socially desirable goals while still delegating considerable autonomy and influence to scientists pursuing reputations for their contributions. Scientific fields have now become highly specialized reputational organizations with varying and changing relations to labour markets, sponsoring agencies, employer policies, and state policies. They are therefore substantially different from their forebears in the seventeenth century and yet share the common attributes of organizing and controlling scientific research through the collective search for collegiate reputations for intellectual contributions to organizational goals. It is how these means of organizing and controlling work through reputations have changed in the past 300 or so years which, I suggest, are key factors in any attempt to understand how and why the modern sciences have developed and altered.

Notes and references

[1] As characterized by H. Martins, 'The Kuhnian " Revolution" and its Implications for Sociology', in T. J. Nossiter *et al.* (eds.), *Imagination and Precision in the Social Sciences*, London: Faber & Faber, 1972.

[2] As exemplified by the National Academy of Sciences report on the Life Sciences. See Committee on Research in the Life Sciences of the Committee on Science and Public Policy of the National Academy of Sciences, *The Life Sciences*, National Academy of Sciences, Washington D. C., 1970, pp. 230–9 and p. 242. The heterogeneity of employment unit titles in the biological sciences is also an instance of the weakening of disciplinary boundaries; see N. Mullins, 'The Distribution of Social and Cultural Properties in Informal Communication Networks among Biological Scientists', *American Sociological Review*, 33 (1968), 786–97.

[3] See B. H. Gustin, *The Emergence of the German Chemical Profession, 1790–1867*, unpublished PhD thesis, University of Chicago, 1975, ch. 5; D. Kevles, 'The Physics, Mathematics and Chemistry Communities: a Comparative Analysis', in A. Oleson and John Voss (eds.), *The Organisation of Knowledge in America, 1869–1920*, John Hopkins University Press, 1979.

[4] As in Warren Weaver's efforts to make biological research more 'scientific' by encouraging 'technology transfer'; see P. Abir-Am, 'The Discourse of Physical Power and Biological Knowledge in the 1930s: a Reappraisal of the Rockefeller Foundation's 'Policy' in Molecular Biology', *Social Studies of Science*, 12 (1982), 341–82.

[5] As Yoxen, among others, has pointed out. See E. Yoxen, 'Life as a Productive Force: Capitalising the Science and Technology of Molecular Biology', in R. M. Young and L. Levidow (eds.), *Studies in the Labour Process*, London: CSE Books, 1981; E. Yoxen, 'Giving Life a New Meaning: the Rise of the Molecular Biology Establishment', in N. Elias *et al.* (eds.), *Scientific Establishments and Hierarchies*, Sociology of the Sciences Yearbook 6, Dordrecht: Reidel, 1982.

[6] See, for instance, élite chemists' claims on behalf of 'little science' and the usefulness of chemistry in: Committee for the Survey of Chemistry, *Chemistry: Opportunities and Needs*, National Academy of Sciences, Washington, D. C., 1965. For a discussion of physicists' and chemists' different priorities in the study of chemical bonding, see D. A. Bantz, 'The Structure of Discovery: Evolution of Structural Accounts of Chemical Bonding', in T. Nickles (ed.), *Scientific Discovery, Case Studies*, Dordrecht: Reidel, 1980.

[7] See, for instance, M. Heirich, 'Why We Avoid the Key Questions: How Shifts in Funding of Scientific Inquiries Affect Decision-Making about Science' in S. Stich and D. Jackson (eds.), *The recombinant DNA Debate*, University of Michigan Press, 1977.

[8] See: F. B. Churchill, 'Chabry, Roux and the Experimental Method in Nineteenth-Century Embryology', in R. N. Giere and R. S. Westfall (eds.), *Foundations of Scientific Method: the Nineteenth Century*, Indiana University Press, 1973; Eugene Cittadino, 'Ecology and the Professionalisation of Botany in America, 1890–1905', *Studies in History of Biology*, 3 (1980), 171–98. On classifications of the sciences in general, see David Knight, *Ordering the World*, London: Deutsch, 1981, pp. 107–52.

[9] B. Latour and S. Woolgar, *Laboratory Life*, London: Sage, 1979, ch. 3.

[10] As evidenced by the studies of laboratories in bio-medical research. See, for example, K. Knorr-Cetina, 'Scientific Communities or Transepistemic Arenas of Research?' *Social Studies of Science*, 12 (1982), 107–30.

[11] See, for instance, M. R. Dando and P. G. Bennett, 'A Kuhnian Crisis in Management Science?', *Journal of the Operational Research Society*, 32 (1981), 91–103; J. W. McGuire, 'Management Theory – Retreat to the Academy', *Business Horizons*, 25 (1982), 31–7; R. D. Whitley, 'The Development of Management Studies as a Fragmented Adhocracy', *Social Science Information*, 23, 1984.

[12] As discussed by N. Elias, 'Scientific Establishments', in N. Elias *et al.* (eds.), *Scientific Establishments and Hierarchies*, Sociology of the Sciences Yearbook 6, Dordrecht: Reidel, 1982.

[13] Which have given rise to intense debates about the status and purpose of literary studies in Britain and the USA. See, for instance, the special issue

on 'Professing Literature' of the *Times Literary Supplement*, 10 December 1982, no. 4, 158 and the subsequent correspondence.

[14] Ibid. It is somewhat ironic that the first field to be 'professionalized' in the German university system was philology, see R. S. Turner, *The Prussian Universities and the Research Imperative, 1806 to 1848*, unpublished PhD thesis, Princeton University, 1972, pp. 292–321. The recent furore over the role of 'theory' in Anglo-Saxon literary studies is probably a reflection of the expansion of numbers and jobs which reduced the existing élite's ability to control access to resources and definitions of skills. Deviant groups have been able to find audiences among their colleagues and develop, at least in the USA, a substantial establishment around 'theory'.

[15] As in the present cut-backs in European universities. The relative power of physics and chemistry departments compared to biology departments in British universities has been highlighted by the implementation of the 1982–4 financial cuts.

[16] As is also demonstrated by the British case where stated government objectives of encouraging university contacts with industry and making research more 'relevant' were translated into the near demolition of the universities with the closest contacts to industry by the use of traditional 'high science' ideals of quality. Similar phenomena occurred in Germany, of course, at the end of the nineteenth century which led to the foundation of the Kaiser-Wilhelm-Gesellschaft and the Physikalisch-Technische Reichsanstalt.

[17] As is shown by the natural scientists' adoption of the humanistic concepts of scholarship and associated institutions after 1835 in Germany. See Turner, op. cit., 1972, note 14, pp. 391–401. Strategic dependence was also heightened by the formation of national associations for science such as the BAAS and the AAAS. For a discussion of how the leaders of the BAAS succeeded in establishing a hierarchy of the sciences in its early years see J. Morrell and A. Thackray, *Gentlemen of Science*, Oxford University Press, 1981, ch. 5.

[18] As charted by Roy Porter, *The Making of Geology*, Cambridge University Press, 1977. See also, W. H. Brock, 'Chemical Geology or Geological Chemistry', and D. E. Allen, 'The Lost Limb: Geology and Natural History', both in L. J. Jordanova and Roy Porter (eds.), *Images of the Earth*, British Society for the History of Sciences Monographs 1, Chalfont St Giles, Bucks, 1979.

[19] Especially in geology and 'natural history', see, for example, D. E. Allen, *The Naturalist in Britain, A Social History*, London: Allen Lane, 1976; P. L. Farber, *The Emergence of Ornithology as a Scientific Discipline: 1760–1850*, Dordrecht: Reidel, 1982, chs. 7 and 8; M. Berman, "Hegemony' and the Amateur Tradition in British Science', *Journal of Social History*, 8 (1975), 30–50; Roy Porter, 'Gentlemen and Geology: the Emergence of a Scientific Career, 1660–1920', *The Historical Journal*, 21 (1978), 809–36. On the role of local, non-professional scientific societies in France in the nineteenth century see R. Fox, 'The *Savant* Confronts His Peers: Scientific Societies in

France, 1815–1914', in R. Fox and G. Weisz (eds.), *The Organization of Science and Technology in France, 1808–1914*, Cambridge University Press, 1980.

[20] By boosting the demand for teachers in departments of physics for instance. See S. Weart, 'The Physics Business in America, 1919–1940', in N. Reingold (ed.), *The Sciences in the American Context*, Washington D. C., Smithsonian Institution Press, 1979; D. Kevles, op. cit., 1979, note 3.

[21] At least to the extent of developing new skills and providing publication space for research in new areas such as geophysics and rubber chemistry.

[22] Hence the Roses' description of the 1914–18 war as 'the chemists' war'. See H. Rose and S. Rose, *Science and Society*, Harmondsworth: Penguin, 1970, ch. 3; M. Sanderson, *The Universities and British Industry 1850–1970*, London: Routledge & Kegan Paul, 1972, ch. 8.

[23] L. R. Harmon and H. Soldz, *Doctorate Production in United States Universities 1920–1962*, Washington D. C.: National Academy of Sciences–National Research Council, 1963, p. 10. Furthermore, this rise continued through the 1930s, especially in chemistry.

[24] Weart, op. cit., 1979, note 20, See also K. Birr, 'Industrial Research Laboratories', in N. Reingold (ed.), *The Sciences in the American Context*, Washington D. C.: Smithsonian Institution, 1979; H. Skolnik and K. M. Reese (eds.), *A Century of Chemistry*, Washington D. C.: American Chemical Society, 1976, pp. 20–4.

[25] Skolnik and Reese (eds.), op. cit., 1976. On the role of the petroleum industry in the expansion of chemical research, see Y. M. Rabkin, 'Chemicalization of petroleum refining in the United States: the role of cooperative research', *Social Science Information*, 19 (1980), 833–50.

[26] P. Forman, *The Environment and Practice of Atomic Physics Weimar Germany: A Study in the History of Science*, unpublished PhD dissertation, University of California at Berkeley, 1967, p. 311. In 1921–2, physics took a half of all the funds allocated for apparatus and 18 per cent of the entire budget. The next largest recipients, chemistry and theology, each had 9 per cent. Within physics, it was the atomic physicists who dominated the funding committees. See also P. Forman, 'Alfred Landé and the Anomalous Zeeman Effect', *Historical Studies in the Physical Sciences*, 2 (1970), 153–261 for an account of the growth of atomic physics after 1918 and the resultant increased competition.

[27] D. Kevles, *The Physicists*, New York: Knopf, 1979, p. 193.

[28] Ibid., pp. 219–20; Stanley Coben, 'The Scientific Establishment and the Transmission of Quantum Mechanics to the United States, 1919–32', *The American Historical Review*, 76 (1971), 442–66.

[29] Over 30 per cent of the Rockefeller Foundation expenditures in natural science between 1933 and 1938 went on what Kohler terms 'Application of Physical Techniques to Biology'. See R. E. Kohler, 'Warren Weaver and the Rockefeller Foundation Program in Molecular Biology: a case study in the management of science', in N. Reingold (ed.), *The Sciences in the American Context*, Washington D. C.: Smithsonian Institution, 1979.

[30] Forman, op. cit., 1967, p. 343, claims that the help provided by the *Notgemeinschaft* to atomic physics was a major factor in its domination of German science.

[31] Not least because of the enormous influence of medical goals and priorities in the legitimation of much biological research.

[32] As documented by Paul Forman, 'Weimar Culture, Causality and Quantum Theory, 1918–1927: Adaptation by German Physicists and Mathematicians to a Hostile Intellectual Environment', *Historical Studies in the Physical Sciences*, 3 (1971), 1–115.

[33] See Weart, 1979, op. cit., note 20, pp. 325–6; Kevles, op. cit., 1979, note 27, ch. 16.

[34] Kevles, ibid., pp. 271–5.

[35] As described by Abir-Am, op. cit., 1982, note 4.

[36] Yoxen, op. cit., 1981, note 5, p. 91.

[37] As suggested by Abir-Am, op. cit., 1982, note 4.

[38] On the homogeneity of university research institutions in Germany and the immense control over research exercised by their heads, see Forman, op. cit., 1967, pp. 96–103.

[39] As in the case of Weaver. See Kohler, op. cit., 1979, note 29; Yoxen, op. cit., 1981, note 5.

[40] As discussed by Coben, op. cit., 1971, note 28 and S. Coben, 'American Foundations as Patrons of Science: the commitment to individual research', in N. Reingold (ed.), *The Sciences in the American Context*, Washington D. C.: Smithsonian Institution, 1979. See also: Kevles, op. cit., 1979, note 27, pp. 197–220 and Weart, op. cit., 1979, note 20, p. 299.

[41] To cite the 'central banker to the world of science', Wickliffe Rose, as quoted by Kevles, op. cit., 1979, note 27, p. 192.

[42] See, for instance, the account by Paul and Shinn of recent French attempts to harness science for economic prosperity: H. W. Paul and T. W. Shinn, 'The Structure and State of Science in France' *Contemporary French Civilisation*, 6 (1981–82), 153–93 at pp. 181–92.

[43] According to Rescher the number of employed natural scientists in the USA rose from 148, 700 in 1950 to 496, 500 in 1970 with the largest percentage increases being recorded by mathematicians, biologists, and medical researchers who all more than quadrupled their numbers in twenty years. See N. Rescher, *Scientific Progress*, Oxford: Blackwell, 1978, p. 59.

[44] Most notably, perhaps by Derek Price. See, for instance, D. J. de Solla Price, *Little Science, Big Science*, New York:Columbia University Press, 1963.

[45] Especially by North American commentators such as W. O. Hagstrom, *The Scientific Community*, New York: Basic Books, 1965, pp. 162–67. See also T. S. Kuhn, 'Second Thoughts on Paradigms', in F. Suppe (ed.), *The Structure of Scientific Theories*, Urbana, University of Illinois Press 1974.

[46] As immortalized by Dan Greenberg, 'Grant Swinger: Reflections on Six Years of Progress', *Science*, 154 (1966), 1424–5.

[47] See, for instance, Yoxen, op. cit., 1981 and 1982, note 5.

[48] See M. Bulmer and J. Bulmer, 'Philanthropy and Social Science in the 1920s: Beardsley Ruml and the Laura Spelman Rockefeller Memorial, 1922–29', *Minerva*, XIX (1981), 347-407 David Morrison, 'Philanthropic Foundations and the Production of Knowledge – a case study', in K. Knorr *et al.* (eds.), *Determinants and Controls of Scientific Development*, Dordrecht: Reidel, 1975.

[49] As in cancer research where many large laboratories support radiobiological, immunological, viral, and pharmacological approaches.

[50] As characterized by Rainer Hohlfeld, 'Two Scientific Establishments which Shape the Pattern of Cancer Research in Germany: Basic Science and Medecine', in N. Elias *et al.*, (eds.), *Scientific Establishments and Hierarchies*, Sociology of the Sciences Yearbook 6, Dordrecht: Reidel, 1982, at p. 164.

[51] For instance, see K. Knorr-Cetina, *The Production of Knowledge*, Oxford: Pergamon, 1981, ch. 4.

[52] See J. J. Salomon, 'Science Policy Studies and the Development of Science Policy', in I. Spiegel-Rösing and D. J. Price (eds.), *Science, Technology and Society*, London: Sage, 1977 for a useful discussion of science policy. A more recent survey of research is S. S. Blume, *Science Policy Research*, Stockholm: Swedish Council for Planning and Coordination of Research, 1981.

[53] In addition to the well-known instances of the Academie Royale des Sciences and the Royal Society, Frederick II of Denmark sponsored perhaps the first example of big science in the sixteenth century – Tycho Brahé's observatory on the island of Ven. See A. Jamison, *National Components of Scientific Knowledge*, University of Lund Research Policy Institute, 1982, pp. 209–25.

[54] See, for example, S. S. Blume, *Toward a Political Sociology of Science*, N. Y.: Free Press, 1974, pp. 180–214; D. Greenberg, *The Politics of American Science*, Harmondsworth: Penguin, 1969, 1969, ch. 8.; M. Callon, 'Struggles and Negotiations to Define What is Problematic and What Is Not', in K. Knorr *et al.* (eds.), *The Social Process of Scientific Investigation*, Sociology of the Sciences Yearbook 4, Dordrecht: Reidel, 1980.

[55] Compare N. Elias, 'Scientific Establishments', in N. Elias *et al.* (eds.), *Scientific Establishments and Hierarchies*, Sociology of the Sciences Yearbook 6, Dordrecht: Reidel, 1982.

[56] Compare M. J. Mulkay, 'The Mediating Role of the Scientific Elite', *Social Studies of Science*, 6 (1976), 445–70.

[57] In the sense of the number of properties considered and the individuality of different arrangements of those properties. Compare C. Pantin, *The Relations between the Sciences*, Cambridge University Press, 1968, ch. 1 and N. Elias, 'The Sciences: towards a theory', in R. Whitley (ed.),

Social Processes of Scientific Development, London, Routledge & Kegan Paul, 1974.

[58] Greenberg, op. cit., 1969, note 54, pp. 148–50.

[59] Yoxen, op. cit., 1981, note 5.

[60] As we found in our study of a cancer research laboratory in the early 1970s. See A. Bitz, A. McAlpine, and R. Whitley, *The Production, Flow and Use of Information in Research Laboratories in Different Sciences*, Machester Business School Research Report Series, 1975, appendix c. Compare, though, Shinn's account of the division of labour and control hierarchy in a French mineral chemistry laboratory: T. Shinn, 'Scientific Disciplines and Organizational Specificity: the Social and Cognitive Configuration of Laboratory Activities', in N. Elias *et al.* (eds.), *Scientific Establishments and Hierarchies*, Sociology of the Sciences Yearbook 6, Dordrecht: Reidel, 1982.

INDEX

Abir-Am, P., 116, 264, 305, 308
Abrams, P., 118, 211, 260, 263
Academic disciplines, 6–7, 56–7, 66, 87, 96, 113,
  decline of monopoly of control over research strategies in 20th Century, 283–9
  distinct from scientific fields, 43–8, 67–8, 82, 113, 163, 278
Academic science, 52–7
Ackoff, R., 260
Aldrich, H. and Mindlin, S., 149
Allen, D. E., 37, 76, 79, 113, 116, 306
Allen, G., 40–1, 80, 114, 151, 209, 210, 215, 216
Anomalies in science,
  and task uncertainty, 129
Artificial Intelligence, 92, 124, 147, 156, 161
  as a professional adhocracy, 191
Art historians, 277
Artis, M., 149, 261
Ash, M., 114, 115, 117, 150, 151, 208, 210, 211, 212, 260, 262
Audience structure, 234–8, diversity of legitimate audiences, 234, equivalence of legitimate audiences, 234–5, combinations of degrees of diversity and equivalence, 235–6
  and employment structures, 236
Augood, D. R., 114

Bachelard, G., 2, 36, 199, 216
Bantz, D. A., 216, 305
Bärmark, J. and Wallen, G., 152
Barnes, S. B. and Dolby, R. G. A., 35
Barnes, S. B. and Shapin, S., 36
Ben-David J., 261
Ben-David, J. and Collins, R., 211
Berman, M., 21, 39, 40, 76, 79, 151, 306
Beyer, J. M. and Lodahl, T. M., 117, 210, 261
Big science and concentration of control, 108, 256–7
  in biology, 299
Biology, as a scientific field:
  compared to physics, 84, 138, 270
  schools in, 196

Biological sciences, 92, 128, 163, 165; 190, 223
  expansion of in USA 291–4, 298
  molecular biology, 194, 286, 287, 292
Biomedical research, 105, 137, 147, 161, 164, 189, 271, 299, 300, contrasted with physics, 5
  expansion of in USA, 250–1, 230
  and National Institutes of Health, 293
Birr, K., 265, 307
Bitz, A., et al., 310
Blaug, M., 213
Bloor, D., 36, 116
Blume, S. S., 35, 261, 309
Blume, S. S., and Sinclair, R., 117, 261, 265
Boas, F., 181, 244
Böhme, G., 150, 212
Böhme, G., et al., 36, 80, 113
Boland, L. A., 213
Bonder, S., 209
Bos, H. J. M., 77
v. d. Braembussche, A., 262
Brock, W. H., 306
Bulmer, M. and Bulmer J., 309
Burns, T. and Stalker, G. M., 148, 150

Caldwell, B., 213
Callon, M., 151, 309
Cancer research, 54, 111, 165, 168, 190, 298
  and cyclotrons, 286
  in West Germany, 294
Caneva, K., 116
Cannon, S. F., 45, 75, 76, 77, 79, 113, 115, 217
Carnegie Institution, 193
Chandler, A. D., 38
Chemistry as a scientific field, 47
  reputational control of research in, 27, 84, 268
  standardized work procedures in, 31, 100
  planning of research in, 70
  contrasted to physics, 89–90, 108, 256–7
  degree of mutual dependence between scientists in, 90–2, 271
  standardization of cognitive objects in, 122

Human sciences (*contd.*)
6 lack of standardized skills in, 26, 186
influence of lay audiences on, 6, 28, 111, 146, 237
subservience to established models of science, 30, 270, 299
low degree of mutual dependence between scientists in, 90–2
and research schools in, 93, 180
use of mathematics in, 106, 282
high degree of task uncertainty in, 127, 135
national and local variations in, 132–3
academic domination of, 164
commonsense objects and language in, 175, 223
as fragmented adhocracies, 168–76, contextual structures of, 240–3
as polycentric oligarchies, 176–81, contextual structures of, 243, 246
limited formalization of, 186–7
plurality of funding sources in, 298
Hutchison, T. W., 212, 213, 214, 260, 262

Industrial science, 51, 53
Intellectual fields as social unit of knowledge production, 7–9
Internal structures of scientific fields, dimensions of, 165–8, degree of specialization and standardization, 166, degree of segmentation, 166–7, degree of differentiation into distinct research schools, 166–7, degree of hierarchical ordering of sub-units, 167, degree of formality of control procedures, 167, degree of theoretical coordination, 167–8, scope and intensity of conflict, 168
summarized for seven major types of science, 169

Jamison, A., 79, 262, 309
Jenkin, P., 149, 185, 213
Johnson, H. G., 151, 210, 213
Johnson, T., 39
Johnston, R. and Jagtenberg, T., 77, 113
Jungnickel, C., 60–1, 76, 79

Kargon, R. H., 39, 79, 262
Karpik, L., 38, 203, 217

Katouzian, H., 117, 149, 212, 213, 214, 262
Kay, N., 38, 213, 214
Kent, R. A., 211, 212, 260
Kevles, D. J., 114, 116, 216, 261, 262, 265, 304, 307, 308
Keylor, W. R., 262
King, M., 35
Knight, D., 305
Knorr, K., 36, 40, 137, 149, 161, 189
Knorr-Cetina, K., 37, 38, 114, 149, 150, 152, 208, 214, 251, 264, 305, 309
Kohler, R., 215, 264, 307, 308
Kuhn, T., 2–5, 35, 36, 38, 65, 80, 114, 119, 129, 149, 164, 184, 209–10, 213, 268, 308
Kuklick, B., 116, 212, 260,
Kuklick, H., 78, 118
Kuper, A., 41, 115, 150, 179, 208, 212, 261, 262, 263

Laboratory sciences:
growth in prestige of, 276, 277, 283–9
Lammers, C., 187, 214
Larson, M. S., 20, 39, 78, 260
Latour, B. and Woolgar, S., 5, 37, 38, 80, 114, 149, 152, 189, 191, 210, 214, 215, 264, 273, 305
Latsis, S., 213
Laudan, L., 35
Law, J., 36
Lawrence, P. R. and Lorsch, J. W., 38
Leamer, E. A., 149
Leipzig psychology, 86, 132
Lemaine, G., 36
Lemert, C., 262
Leontief, W., 208, 212, 214
Levantman, S., 211, 212
Liebig, J. V., 60, 62, 100, 107, 211, 301
Lighthill, Sir J., 215
Literary studies as a scientific field, 132, 133, 234, 243
professionalism in, 278
Loasby, B. J., 214
Lodahl, J. B. and Gordon, G., 208
Lundgreen, P., 211, 261, 264, 265

Machlup, F., 213
Malinowski, B. C., 179
Management studies as a scientific field, 91, 141, 158, 163, 164, 167, 173, 221, 222–3, 241, 243, 273
Manegold, K. H., 78

Mannheim, K., 35
Marchi, N. de, 214
Martin, B., 116, 210
Martins, H., 35–6, 187, 214, 304
Mathematics, US, 91, 92, 138, 158 as
  professional adhocracy, 191–2
  specialization in, 192
  manifesting particular degrees of
  reputational autonomy and control, 226–7
  national traditions in 252–4
  in Germany, 44
Mauskopf, S. and McVaugh, M. R., 260
Mehrtens, H., 76
Mendelsohn, E., 37, 78
Merton, R., 38
Mintzberg, H., 113, 149, 159, 208, 216,
  227, 261
Modern sciences, essential characteristics of, 10–29, as systems of
  novelty production, 11–12, tension between innovation and
  tradition in, 13, as a type of craft
  administration system, 14–19, as
  a type of professional administration system, 19–25, as reputational systems of work organization and control, 25–9
  conditions for their establishment as
  distinct systems of work organization and control, 29–32
  and professional self-interest, 139–
  40
Morrell, J., 41, 60, 76, 79, 116, 117
Morrell, J., and Thackray, A., 38, 39, 76,
  77, 113, 306
Morrison, D., 309
Moseley, R., 79
Moss, S., 213
Mulkay, M., 35, 309
Mulkay, M., and Edge, D., 116
Mullins, N., 115, 117, 150, 209, 210, 214,
  215, 245, 263, 304
Mutual dependence between scientists,
  degree, of, 87–95
  functional, 88, strategic, 88–9, and
  competition, 88, 94, 96–7 low
  degrees of, 90–2, high functional
  and low strategic, 92, low functional and high strategic, 92–3,
  high degrees of, 93–4, interrela-
  tions with the degree of task
  uncertainty, 153–9

and the organization of scientific
  work, 95–104
and contextual factors, 104–12
Mutual dependence between scientific
  fields, 267–71, functional dependence, 268–9, strategic dependence, 269, interrelation between types of dependence between fields, 269–71
  increasing dependence between fields
  and changes in their organization, 271–6, and increasing specialization, 272, declining disciplinary autonomy, 272, and increasing competition between
  fields, 273–4, and theoretical
  coordination of knowledge across
  fields, 275
McCann, H. Gilman, 38, 78
McClelland, C. E., 38, 78, 79, 80, 211,
  261
McCloskey, D. M., 213, 214
McGuire, J. W., 305
McIntosh, R. P., 211
Mackenzie, B. D., 115, 151, 210

Nelson, R. and Winter, S. G., 208, 212,
  214
Nickles, T., 35
Normalization of research in professionalized science, 60

O'Connor, J. G. and Meadows, A. J., 76,
  79
Ornithology, 100, 124, 158
Outram, D., 78

Pantin, C. F. A., 38, 214, 217, 309
Parsons, T., 175
Partitioned bureaucracy as a type of
  ·scientific field, 160
  internal structure of, 181–7, core and
  periphery differentiated, 181–3
  contextual features of 246–9, varia-
  tions between core and periphery,
  247
Paul, H. W., 216
Paul, H. W., and Shinn, T., 308
Perrow, C., 120, 122, 149
Pfetsch, F., 36
Philosophy as a scientific field, 91, 132,
  158, changes in, 99
Philosophy of science, 1–2

as a type of craft work organization, 14–19
as a type of professional work organization, 19–25
as a reputational work organization, 25–9
conditions for their establishment, 29–32
comparative analysis of, 83–7
reflecting differing degrees of functional and strategic dependence between scientists, 90–5
reflecting differing degrees of technical and strategic task uncertainty, 124–30
reflecting differing combinations of mutual dependence and task uncertainty, 153–9
seven major types of, 158–64
internal structures of, 164–205
mutual dependence between, 267–76
Scientific research as a type of craft work, 10, 14–19
as coordinated novelty production, 11–13, 17–19, 27
as uncertainty reduction, 138–41
Shapin, S. and Thackray, A., 35
Shinn, T., 79, 80, 115, 199–200, 209, 216, 228–9, 261, 264, 310
Shubik, M., 213
Silliman, R. H., 77, 217
Simpson, R. L., 211
Size of scientific fields, and competition, 109
and standardization, 109
Skolnik, H., and Reese, K. M., 307
Social anthropology, 31, 180–1
graduate training in, 245
British, 93, 158, 179, 221, 229, 244
Sociology as a scientific field, anglosaxon, 91, 106, 124, 134, 136, 143, 163, 165, 167, 168, 175, 176, 238, 241
and control over graduate education, 244–5
British, 222, 224, 243
Sociology of science, 1–6
European, 3–6
Sommerfeld, A., 107
Specialization in the sciences, 28, 30, 62, 166–7
and the degree of functional dependence between scientists, 94, 96

and size of field, 109
in fragmented adhocracies, 168, 175
in polycentric oligarchies, 178
in partitioned bureaucracies, 182
in professional adhocracies, 188, 190
in polycentric professions, 193–4
in technologically integrated bureaucracies, 198–9
and changes in the organization and control of research, 295
and state science policies, 300
Sprague, L. G. and Sprague, C. R., 260
Standardization of techniques and work procedures, 31, 65–6, 166–7
and task uncertainty, 120–2, 125, 130–1
and reputational autonomy, 142–3
and concentration of control, 143–5
and audience diversity, 146
and changes in the organization and control of research, 295
and state science policies, 276, 298, 300
Standardized symbol systems, 31–2
and increasing mutual dependence, 98
and technical task uncertainty, 134–5
and patronage, 144
and audience diversity, 146.
Stanfield, J. R., 117
Starbuck, W. H., 117
Starnberg group, 3, 36
State science, 51–7
State science policies, development of, 295–301
and sociology of science, 2
and formation of a trans-scientific establishment, 297–8
consequences for the organization of scientific fields, 298–300
Stigler, G., 212, 213
and Friedland, C., 263
Stinchcombe, A., 15–18, 27, 38, 39, 149, 150, 217
Stocking, G. W., 212, 263
Storer N., 8, 37, 210
Strategic dependence between scientists, degree of, 88–9
in Physics, 89–90
in Chemistry, 90
and competition over goals, 101–3
and theoretical coordination of research, 102–3

Wilson, D. J., 150
Woodward, J., 148, 204, 217, 218
Wundt, W., 86, 107, 178–9, 211
Würzburg psychology, 86, 132, 136

Wynne, B., 37

Yoxen, E., 78, 80, 116, 215, 216, 260,
264, 287, 305, 308, 309, 310